How to Pass

ADVANCED HIGHER

Biology

Billy Dickson
and Graham Moffat

HODDER
GIBSON
AN HACHETTE UK COMPANY

The Publishers would like to thank the following for permission to reproduce copyright material.

Photo credits

p.11 (left) © sinhyu/stock.adobe.com (centre) al7/Shutterstock.com; **p.13** © StockPhotoPro/stock.adobe.com; **p.15** © StockPhotoPro/stock.adobe.com; **p.51** © Heiti Paves/Shutterstock.com; **p.66** (top right) © alba1988/stock.adobe.com; (bottom left) © bravikvl/Shutterstock.com; (bottom right) © Dennis W Donohue/Shutterstock.com; **p.67** © Cara Siera/Shutterstock.com; **p.68** © Jesus/stock.adobe.com; **p.70** (left and right) © Graham Moffat; **p.76** (top right) © Graham Moffat; (centre left) © Michele Aldeghi/Shutterstock.com; (centre right) © Sergey Uryadnikov/Shutterstock.com; **p.77** (left) © Steve McWilliam/Shutterstock.com; (right) © IanRedding/Shutterstock.com; **p.79** © adityajati1995/stock.adobe.com; **p.80** top © KeikoH/Shutterstock.com; (bottom) © lesniewski/stock.adobe.com; **p.86** (left) © Kazakova Maryia/stock.adobe.com; (right) © Wirestock/stock.adobe.com; **p.90** (top left) © David Thyberg/stock.adobe.com; (top centre) © Sriyana/Shutterstock.com; (top right) © Sid10/stock.adobe.com; (centre right) © jmagtec/stock.adobe.com

Acknowledgements

Every effort has been made to trace all copyright holders, but if any have been inadvertently overlooked, the Publishers will be pleased to make the necessary arrangements at the first opportunity.

Although every effort has been made to ensure that website addresses are correct at time of going to press, Hodder Gibson cannot be held responsible for the content of any website mentioned in this book. It is sometimes possible to find a relocated web page by typing in the address of the home page for a website in the URL window of your browser.

Hachette UK's policy is to use papers that are natural, renewable and recyclable products and made from wood grown in well-managed forests and other controlled sources. The logging and manufacturing processes are expected to conform to the environmental regulations of the country of origin.

Orders: please contact Hachette UK Distribution, Hely Hutchinson Centre, Milton Road, Didcot, Oxfordshire, OX11 7HH. Telephone: +44 (0)1235 827827. Email education@hachette.co.uk. Lines are open from 9 a.m. to 5 p.m., Monday to Friday. You can also order through our website: www.hoddereducation.co.uk. If you have queries or questions that aren't about an order, you can contact us at hoddergibson@hodder.co.uk

First published in 2021 by
Hodder Gibson, an imprint of Hodder Education
An Hachette UK Company
50 Frederick Street
Edinburgh, EH2 1EX

Impression number   6   5

Year   2025 2024

Cover photo © uwimages/stock.adobe.com
Typeset in India by Aptara Inc.
Printed by CPI Group (UK) Ltd, Croydon CR0 4YY

A catalogue record for this title is available from the British Library.

ISBN: 978 1 3983 1217 3

SCOTLAND EXCEL

We are an approved supplier on the Scotland Excel framework.

Find us on your school's procurement system as:

Hachette UK Distribution Ltd or *Hodder & Stoughton Limited t/a Hodder Education.*

MIX
Paper | Supporting responsible forestry
FSC™ C104740
www.fsc.org

Contents

Introduction

Welcome to *How to Pass SQA Advanced Higher Biology*. The Advanced Higher Biology course is split into three areas of study – cells and proteins, organisms and evolution, and investigative biology – and an individual project.

How to use this book

The knowledge content is covered in Key Areas 1.1 to 1.5 and 2.1 to 2.5. The skills content is largely covered in Key Areas 3.1 to 3.3.

Key points

Each Key Area starts with a list of success criteria called key points. Each point has a check box into which you can traffic light your confidence level for that information: green for 'confident', amber for 'needs more work' and red for 'need teacher help'.

Summary notes

The main text provides concise, illustrated expansions of the key points. These should be read carefully several times and should be revisited throughout the course. Hints and tips are given in the margins, while key links cross-reference to ideas that appear in other areas of the book – you should follow these.

Check-up questions

The text is split into handy chunks by numbered check-up questions. These should be answered as you progress through a key area and self-marked to provide feedback to help in traffic lighting the key points. Answers are given at the end of the book.

Key words

Scoring marks in SQA exams requires knowledge of technical terms – these will appear regularly throughout the question paper and, most importantly, its marking instructions. The key words appear in **bold** in the key points and are further defined in key word boxes. You should study the key words carefully. We recommend that you make a set of flashcards for each key area – index cards with the term on one side and the definition on the other.

Exam-style questions

After each Key Area are some SQA-style structured and extended-response exam-style questions that bring together many of the ideas in the Key Area. These should be completed soon after finishing the study of a Key Area. It is probably better to self-mark these and use your score as feedback for revision of your weaker areas.

Answers

These provide easily accessible and understandable answers to all check-up and exam-style questions, and give mark breakdowns.

Practice course assessment

At the end of each of the three areas is a test with a sample of questions that will allow you to evaluate your overall progress in that area and provide feedback for your revision. The tests for Areas 1 and 2 are for 50 marks and you could mark and grade your work:

- 25 for a C pass
- 30 for a B pass
- 35 for an A pass.

The Area 3 test is for 30 marks:

- 15 for a C pass
- 18 for a B pass
- 21 for an A pass.

Practice course assessment answers are given on pages 152, 158–159 and 163–164.

Your project and exam

The *Your project* section on pages 138–141 has more information and helpful hints for your project. Part 4 has more information and helpful hints for your exam.

WWW

- SQA course specification, past papers and project information: **www.sqa.org.uk**
- SQA's examples of candidate evidence with marks and commentaries: **www.understandingstandards.org.uk**
- Scottish Schools Education Research Centre (SSERC) additional project support: **www.sserc.org.uk**
- Information on course assessment and grading, general exam advice and a downloadable record of progress and evaluation: **www.hoddergibson.co.uk/ah-biology-extras**

Area 1 Cells and proteins

Laboratory techniques for biologists

Key points !

1 Substances, organisms and equipment in a laboratory can present a **hazard**. ☐
2 Hazards in the lab include toxic and corrosive chemicals, heat and flammable substances, **pathogenic** organisms, and mechanical equipment. ☐
3 Risk is the likelihood of harm arising from exposure to a hazard. ☐
4 Risk assessment involves identifying possible risks and the control measures to minimise them. ☐
5 Control measures used to minimise risk include using appropriate handling techniques, protective clothing and equipment, and **aseptic technique**. ☐
6 Dilutions in a **linear dilution series** differ by an equal interval, for example 0.1 M, 0.2 M, 0.3 M, and so on. ☐
7 Dilutions in a **log dilution series** differ by a constant proportion, for example 10^{-1}, 10^{-2}, 10^{-3}, and so on. ☐
8 A **standard curve** is produced by plotting measured values for known concentrations; it is used to determine the concentration of an unknown solution. ☐
9 **Buffers** are used to control pH; the addition of acid or alkali only has a very small effect on its pH, allowing the pH of a reaction mixture to be kept constant. ☐
10 A **colorimeter** can be used to quantify the concentration and **turbidity** of a solution. The colorimeter is calibrated using an appropriate blank as a baseline. The measurement of absorbance is used to determine the concentration of a coloured solution using suitable wavelength filters. The measurement of percentage transmission is used to determine turbidity. ☐
11 **Centrifugation** is a technique used to separate substances of differing density. More dense components settle in a pellet; less-dense components remain in the **supernatant**. ☐
12 Paper and thin layer **chromatography** can be used for separating different substances such as amino acids and sugars. ☐
13 The speed that each solute travels along the chromatogram depends on its solubility in the solvent used. ☐
14 **Affinity chromatography** is a separation technique in which soluble target proteins with a high **affinity** in a mixture become attached to specific molecules as the mixture passes down a column. Non-target molecules with a weaker affinity are washed out. ☐
15 **Gel electrophoresis** is a technique that can be used to separate proteins and nucleic acids. ☐
16 In gel electrophoresis, charged macromolecules move though an electric field applied to a gel matrix. ☐
17 **Native gel electrophoresis** separates proteins and nucleic acids by their shape, size and charge. ☐
18 Native gels do not denature the molecules, so the separation is by shape, size and charge. ☐
19 **SDS–PAGE** gives all the molecules an equally negative charge and denatures them, separating proteins by size alone. ☐
20 Proteins can be separated from a mixture using their **isoelectric points (IEPs)**. ☐

⇨

⇨

21 The IEP is the pH at which a soluble protein has no net charge and will precipitate out of solution.   ☐

22 If the solution is buffered to a specific pH, only the protein(s) that have an IEP of that pH will precipitate.   ☐

23 Soluble proteins can be separated using an electric field and a pH **gradient gel**.   ☐

24 A protein stops migrating through the gel at its IEP in the pH gradient because it has no net charge.   ☐

25 Proteins can be detected using antibodies.   ☐

26 **Immunoassay** techniques are used to detect and identify specific proteins. These techniques use stocks of antibodies with the same specificity, known as **monoclonal antibodies**.   ☐

27 An antibody specific to the protein **antigen** is linked to a chemical 'label'.   ☐

28 The label is often a **reporter enzyme** producing a colour change, but chemiluminescence, fluorescence radioactivity and other reporters can be used.   ☐

29 In some cases, the assay uses a specific antigen to detect the presence of antibodies.   ☐

30 **Western blotting** is a technique used after SDS–PAGE electrophoresis. In western blotting, the separated proteins from the gel are transferred (blotted) on to a solid medium.   ☐

31 The proteins can be identified using specific antibodies that have reporter enzymes attached.   ☐

32 **Bright-field microscopy** is commonly used to observe whole organisms, parts of organisms, thin sections of dissected tissue or individual cells.   ☐

33 **Fluorescence microscopy** uses specific **fluorescent** labels to bind to and visualise certain molecules or structures within cells or tissues.   ☐

34 Aseptic technique eliminates unwanted microbial contaminants when culturing micro-organisms or cells.   ☐

35 Aseptic technique involves the sterilisation of equipment and culture media by heat or chemical means, and subsequent exclusion of microbial contaminants.   ☐

36 A microbial culture can be started using an **inoculum** of microbial cells on an agar medium, or in a broth with suitable nutrients.   ☐

37 Many **culture media** exist that promote the growth of specific types of cells and microbes.   ☐

38 Animal cells are grown in medium containing **growth factors** from **serum**.   ☐

39 Growth factors are proteins that promote cell growth and proliferation.   ☐

40 Growth factors are essential for the culture of most animal cells. In culture, **primary cell lines** can divide a limited number of times, whereas tumour cell lines can perform unlimited divisions.   ☐

41 Plating out of a liquid microbial culture on solid media allows the number of colony-forming units to be counted and the density of cells in the culture estimated.   ☐

42 Serial dilution is often needed to achieve a suitable colony count.   ☐

43 **Haemocytometers** are used to estimate cell numbers in a liquid culture.   ☐

44 **Vital staining** is required to identify and count viable cells.   ☐

Health and safety

Hazards

Biology laboratories often have substances, organisms and equipment that can present hazards to the people that work there. Figure 1.1 shows some of the main hazard warning signs you may see in biology laboratories.

Toxic chemicals

Toxic chemicals are substances that are harmful when inhaled, ingested, injected or absorbed. Essentially, they are poisons. Some are inorganic and others are produced by living organisms. The degree of toxicity depends on the concentration of the substance. Substances that are toxic to humans may not necessarily harm other species.

Corrosive chemicals

A corrosive chemical is a reactive substance that can damage living tissue. They act either directly, by chemically destroying part of the body, or indirectly, by causing inflammation. Acids and bases are corrosive substances commonly found in biology laboratories.

Heat and flammable substances

Sources of heat that laboratory workers should be aware of to prevent burns and scalds include lighted Bunsen burners, electric ovens, hotplates and steam baths. Flammable substances are those that can be easily ignited at room temperature.

Pathogenic organisms

Pathogens are organisms that can cause disease. They are a biohazard.

Mechanical equipment

Mechanical equipment includes machines that may have moving or vibrating parts, hot surfaces, or sharp or heavy components. The main physical risks from mechanical equipment include noise; hand, arm and whole-body vibration; and heat stress. The main mechanical risks include cuts, lacerations, needle punctures and crushing. Different hazard symbols warn of the specific danger posed by mechanical equipment.

Figure 1.1 Some of the main hazard warning signs: a) toxic, b) corrosive, c) flammable, d) biohazard, e) and f) specific mechanical warning signs

Risk and risk assessment

Risk is the likelihood of harm arising from exposure to a hazard. Risk assessment involves identifying risk levels, their likely severity and the control measures that can be used to minimise these risks. Control measures include:

- using appropriate handling and disposal techniques
- using appropriate masses, volumes and concentrations of substances
- use of protective clothing, such as laboratory coats and gloves
- use of protective equipment, such as goggles and masks
- use of aseptic technique in microbiology.

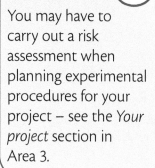

Key links

You may have to carry out a risk assessment when planning experimental procedures for your project – see the *Your project* section in Area 3.

A grid for carrying out a risk assessment is shown in Figure 1.2.

Hazard identified	Risk level (low, medium, high)	Severity of risk (low, medium, high)	Control measures to be used (handling techniques, equipment and clothing, concentrations of substances, disposal methods, etc.)

Figure 1.2 Headers of a risk assessment grid – there are many more examples of control measures

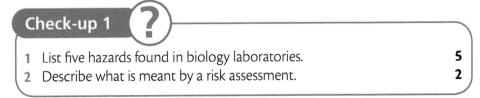

Check-up 1

1   List five hazards found in biology laboratories.   **5**
2   Describe what is meant by a risk assessment.   **2**

Liquids and solutions
Making and using linear and log dilutions

Linear dilution is often used when the substance being diluted is the independent variable in an experiment. Linear dilutions differ from each other by an equal interval. To make a linear dilution of a substance, for example glucose, start with a stock solution of a known concentration of glucose in the solvent (distilled water). Add increasing, stepped volumes of that solution to separate tubes, then add pure solvent (distilled water) to each tube so that an equal volume of each dilution is produced, as shown in Figure 1.3.

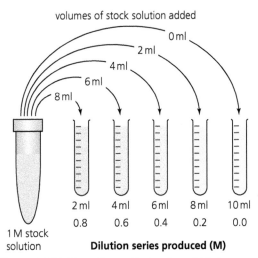

volumes of stock solution added

0 ml
2 ml
4 ml
6 ml
8 ml

1 M stock solution

	2 ml	4 ml	6 ml	8 ml	10 ml
	0.8	0.6	0.4	0.2	0.0

Dilution series produced (M)

Figure 1.3 Making a linear dilution from 1 M glucose solution – various volumes of the stock are added to separate tubes and diluted up to the same volume with solvent (distilled water) to produce the linear dilutions from 0.0 M to 0.8 M

Log dilution is often used in microbiology to estimate the concentration or density of cells in a stock culture. Dilutions in a log dilution series differ by a constant proportion, for example 10^{-1}, 10^{-2}, 10^{-3}. They are created by diluting a stock solution by a factor then further diluting the dilution produced by the same factor, and so on, as shown in Figure 1.4. This serial dilution produces low concentrations of cells that can be cultured on an agar plate, producing a number of easily countable colonies. From this result, the estimated number of cells in the original stock can be calculated.

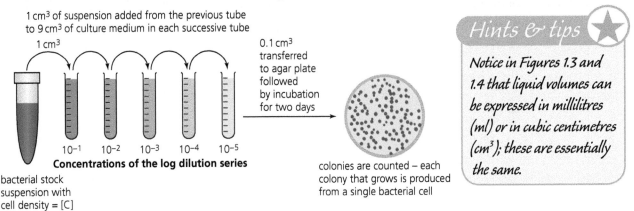

$1 cm^3$ of suspension added from the previous tube to $9 cm^3$ of culture medium in each successive tube

$1 cm^3$

$0.1 cm^3$ transferred to agar plate followed by incubation for two days

10^{-1} 10^{-2} 10^{-3} 10^{-4} 10^{-5}
Concentrations of the log dilution series

bacterial stock suspension with cell density = [C]

colonies are counted – each colony that grows is produced from a single bacterial cell

Hints & tips

Notice in Figures 1.3 and 1.4 that liquid volumes can be expressed in millilitres (ml) or in cubic centimetres (cm^3); these are essentially the same.

Figure 1.4 Making a log dilution from a cell culture – each individual bacterium grows to a colony that can be counted, then a calculation can be carried out to give a value for bacteria per unit of culture volume (the cell density)

Colorimetry

A colorimeter can be used to determine the concentration of solutions that have been coloured by an indicator reagent by measuring how much light they absorb. Light is passed through a sample of the solution contained in a small tube called a cuvette; an electronic sensor detects how much light has been absorbed as it passed through the solution or culture suspension, as shown in Figure 1.5. A typical school colorimeter is also shown.

Colorimeters can also be used to determine the turbidity of a cell culture. Turbidity is proportional to the density of cells in the culture. Light is passed through a sample of the culture in a cuvette; an electronic sensor detects how much of light has been transmitted through the culture suspension.

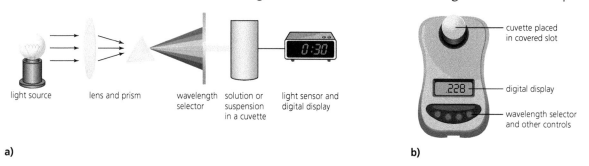

a)

light source   lens and prism   wavelength   solution or   light sensor and
selector   suspension   digital display
in a cuvette

cuvette placed
in covered slot

digital display

wavelength selector
and other controls

b)

Figure 1.5 a) Components of a colorimeter; b) a typical school colorimeter with controls to set the device to read either absorbance or transmission of specific wavelengths

Before measuring concentration or turbidity with a colorimeter, the instrument must be calibrated using a blank as a baseline. A blank is a cuvette containing only the solvent used when making the dilutions or a sample of the medium used in the cell culture.

Standard curves

A linear dilution series of a substance such as glucose can be used to produce a standard curve. Each dilution of a glucose dilution series has a reagent added that reacts with the glucose to give a coloured product; the colour is proportional to the glucose concentration. The absorbance of the different coloured glucose solutions can then be measured with a colorimeter and the results used to plot a standard curve, as shown in Figure 1.6a).

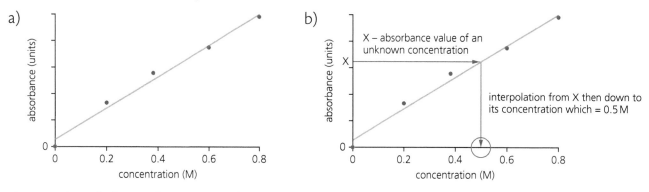

a)

absorbance (units)

0
0   0.2   0.4   0.6   0.8
concentration (M)

b)

absorbance (units)

X – absorbance value of an
unknown concentration

X

interpolation from X then down to
its concentration which = 0.5 M

0
0   0.2   0.4   0.6   0.8
concentration (M)

Figure 1.6 a) Standard curve of concentration of a glucose solution; b) using the curve to find the concentration of a glucose solution of unknown concentration

A glucose solution of unknown concentration can be added to the same reagent to produce a coloured solution; its concentration can then be determined by interpolation to the standard curve, as shown in Figure 1.6b).

This general method can also be used to determine unknown cell culture densities.

Use of buffers to control pH

Buffers are solutions that can resist changes of pH even although acid or alkali is added. This allows the pH of a reaction mixture to be kept constant in spite of the production of acidic or alkaline products.

Most biological reactions are dependent on pH, so buffer solutions are often used in *in vitro* experiments on these reactions so that changes in pH during the reaction don't act as confounding variables and cause a mistaken association between the independent and dependent variables.

Check-up 3

1 Describe how you could use colorimetry and a standard curve to identify the glucose concentration of an unknown solution.   **4**
2 Explain what buffers are and why they are used in experiments.   **2**

Key links

There is more about confounding variables and *in vitro* procedures in Key Area 3.2a.

Separation techniques

Centrifugation to separate substances of differing density

Centrifugation is used to separate components of a suspension that have a different density. For example, different components of cells can be separated by homogenisation of tissue followed by centrifugation. The cell homogenate is placed in a centrifuge tube, which is then spun in a centrifuge machine at between 200 and 120,000 revolutions per minute (rpm). After a time, the denser components of the cells are separated into a pellet while less-dense components remain suspended in the supernatant.

The homogenisation of potato cells and the separation of starch grains from enzyme-rich cytoplasm is shown in Figure 1.7.

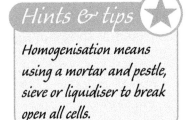

Hints & tips

Homogenisation means using a mortar and pestle, sieve or liquidiser to break open all cells.

a)

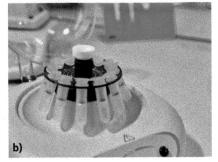
b)

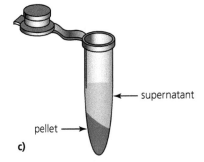
← supernatant

pellet →

c)

Figure 1.7 a) Homogenisation of potato cells using a mortar and pestle; b) spinning homogenate in a centrifuge; c) the separation of dense starch grains in a pellet from less-dense cytoplasm in the supernatant

Paper and thin layer chromatography

Chromatography can be used to separate different solutes such as amino acids and sugars. Mixtures of these substances dissolved in a solvent can be added to a paper strip or to a metal foil strip with a thin layer of silica or cellulose bonded to it. The speed that each solute travels along the strip depends on its differing solubility in the chromatography solvent used, and its differing affinity for the paper or thin layer. If the substances being separated are colourless, like amino acids, they must be made visible on the paper or thin layer using a developing agent, as shown in Figure 1.8.

Hints & tips

The solubility and affinity of an amino acid is dependent on its R group.

Key links

There is much more about R groups in Key Area 1.2.

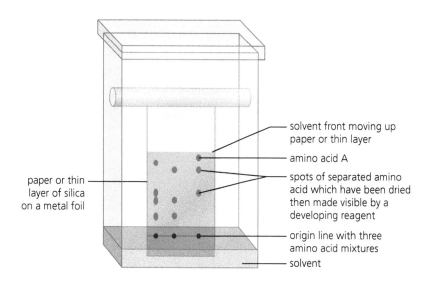

paper or thin layer of silica on a metal foil

solvent front moving up paper or thin layer

amino acid A

spots of separated amino acid which have been dried then made visible by a developing reagent

origin line with three amino acid mixtures

solvent

Figure 1.8 Separation of amino acids by paper or thin layer chromatography – amino acid A travels the furthest up the paper because of the properties of its R group, which determine its solubility and affinities

Affinity chromatography

Affinity chromatography can be used to separate target proteins from a mixture of proteins. A solid gel column in a glass tube is produced with specific molecules, such as antibodies or ligands, bound to the gel. Soluble target proteins with a high affinity for these specific molecules become attached to them as a mixture of proteins passes down the column. Non-target proteins with a weaker affinity or no affinity are washed out. The target protein can then be washed out separately and collected, as shown in Figure 1.9.

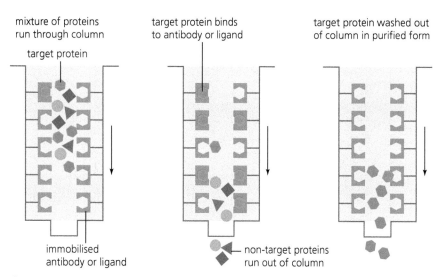

mixture of proteins run through column

target protein

target protein binds to antibody or ligand

target protein washed out of column in purified form

immobilised antibody or ligand

non-target proteins run out of column

Figure 1.9 Separating a target protein by affinity chromatography

Check-up 4

1 Explain how centrifugation achieves separation of a suspension into a pellet and a supernatant. **2**
2 Describe how a solution with three different amino acids can be separated. **3**
3 Describe how proteins in a mixture can be separated by affinity chromatography. **3**

Key links 👍

There is more about ligands in Key Area 1.2, and more about antibodies in Key Area 2.5.

Gel electrophoresis (PAGE)

In electrophoresis charged macromolecules, such as proteins or nucleic acids, move through an electric field applied to a buffered gel matrix (polyacrylamide gel electrophoresis, or PAGE) as shown in Figure 1.10.

Native PAGE gels do not denature the molecules being separated – they preserve their structure and function – so the separation is by shape, size and charge, but it is tricky to carry out.

SDS–PAGE gels contain sodium dodecyl sulfate (SDS), which denatures molecules passing through it. It gives all the molecules present an equally negative charge, separating proteins by size alone. It is simple to carry out, but the structure and function of any separated protein is lost.

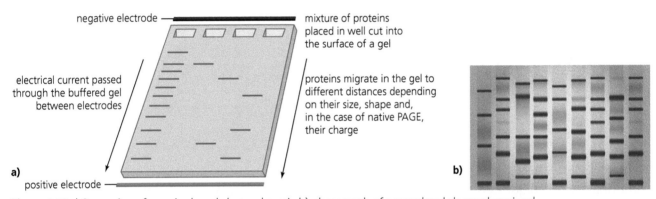

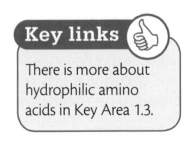

Figure 1.10 a) Separation of proteins by gel electrophoresis; b) photograph of a completed electrophoresis gel

Isoelectric point (IEP)

Proteins have net charges caused by the R groups of their hydrophilic surface amino acids. The isoelectric point (IEP) is the pH value at which a protein is electrically neutral, with surface charges that are balanced as shown in Figure 1.11. At pHs below the IEP the net charges are positive, and at pHs above the IEP the net charges are negative, allowing the protein to remain in solution or suspension. At IEP, the protein loses solubility in water and starts to solidify and precipitate out of solution.

Key links 👍

There is more about hydrophilic amino acids in Key Area 1.3.

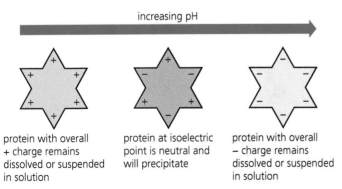

Figure 1.11 Protein molecule with surface charges, which change depending on the pH of its surroundings – at IEP there is no net charge and the protein precipitates

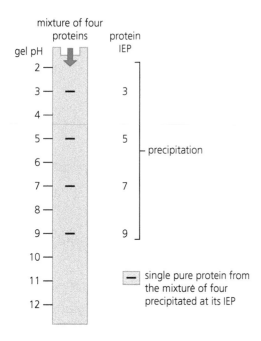

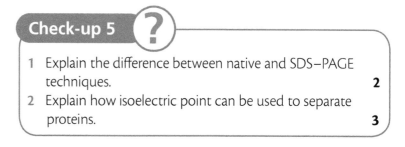

IEP can be used to separate a mixture of proteins by a type of electrophoresis in which the mixture is passed through a gel with a built-in pH gradient. The individual proteins reach their IEPs one by one and are collected as precipitates, as shown in Figure 1.12.

Check-up 5

1 Explain the difference between native and SDS–PAGE techniques. **2**
2 Explain how isoelectric point can be used to separate proteins. **3**

Figure 1.12 Four proteins of different isoelectric point being separated in a pH gradient gel electrophoresis procedure

Detecting and identifying proteins using antibodies

Immunoassay techniques

Immunoassay techniques are used to detect and identify specific proteins. These techniques use stocks of antibodies with the same specificity, known as monoclonal antibodies. The antibodies can be linked to a chemical 'label', often a reporter enzyme that produces a colour change, but reporters that show by chemiluminescence, fluorescence or radioactivity can also be used.

In the simple example in Figure 1.13, a sample to be assayed has been added to a container and antigenic material adheres to the container surface. To identify if a specific antigen is present, a labelled antibody specific to the antigen to be detected or identified is added and then washed out. In the case of a reporter enzyme label, its colourless substate is then added, which will change to a coloured product if the enzyme has been trapped on the antibody bound to the antigen.

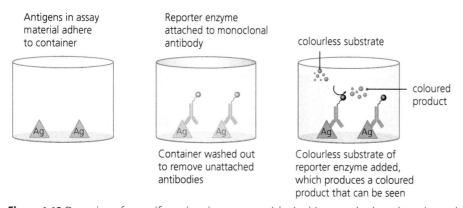

Figure 1.13 Detection of a specific antigen in assay material – in this example, the colour change happens if the reporter is bound, and it can only be bound if the specific antigen is present

Other immunoassay techniques involve the antibody being pre-attached to the container and the sample being analysed added. If the antigen is present in the sample, it will specifically bind to the antibody in the container and can't be washed away. A second antibody with a reporter that binds to the first can then be added and the reporter signal looked for.

Western blotting

Western blotting is a technique used after SDS–PAGE electrophoresis. The separated proteins from the gel are transferred (blotted) on to a solid medium or membrane and dried. The proteins can be labelled by soaking the blot with specific antibodies. The antibodies bind to specific proteins. A second antibody with reporter enzymes – radioactive tags or fluorescent markers attached – are then added. The presence of the protein can then be visualised by eye following the addition of the reporter enzyme's substrate, which changes colour, or by seeing fluorescence in UV light or by using photographic paper, which is sensitive to radioactivity, as shown in Figure 1.14.

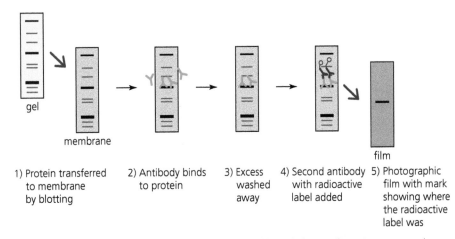

gel

membrane

film

1) Protein transferred to membrane by blotting

2) Antibody binds to protein

3) Excess washed away

4) Second antibody with radioactive label added

5) Photographic film with mark showing where the radioactive label was

Figure 1.14 Western blotting – 1) proteins separated by gel electrophoresis are pressed on to a membrane; 2) membrane is treated with antibodies specific to a target protein; 3) excess is washed away but some antibody binds specifically to target; 4) a second antibody with a radioactive reporter is added and binds to any antibody already present; 5) membrane placed against a piece of photographic film – the radioactivity leaves a mark showing the presence of the target protein

Microscopy

Bright-field microscopy is used to examine whole organisms, parts of organisms, thin sections of dissected tissue or individual cells. Figure 1.15 shows some onion cells undergoing cell division by mitosis. The cells have been stained with a pink stain to make them more visible.

Fluorescence microscopy uses UV light to detect specific fluorescent stains, which bind to and visualise certain molecules or structures within cells or tissues.

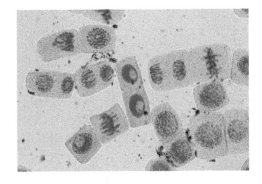

Figure 1.15 Stained onion cells undergoing mitosis captured using bright-field microscopy

Key links 👍

An example of an image captured using fluorescence microscopy is shown in Figure 1.47a) on page 51.

Key links 👍

There is more about mitosis in Key Area 1.5.

Aseptic technique and cell culture

Aseptic technique eliminates unwanted microbial contaminants when culturing micro-organisms or cells. Aseptic technique involves the sterilisation of equipment and culture media by heat or chemical means, and the subsequent exclusion of microbial contaminants, as shown in Figure 1.16.

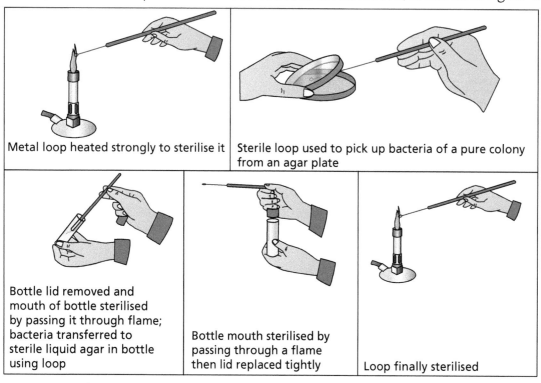

Metal loop heated strongly to sterilise it	Sterile loop used to pick up bacteria of a pure colony from an agar plate
Bottle lid removed and mouth of bottle sterilised by passing it through flame; bacteria transferred to sterile liquid agar in bottle using loop	Bottle mouth sterilised by passing through a flame then lid replaced tightly Loop finally sterilised

Figure 1.16 Simple example of the use of aseptic transfer technique in a biology laboratory – an inoculum is taken from a sterile culture growing on agar in a Petri dish using an inoculating loop, added to sterile liquid agar in a labelled culture bottle and sealed

A microbial culture can be started using an inoculum of microbial cells on an agar medium, or in a broth with suitable nutrients.

Many culture media exist that promote the growth of specific types of cells and microbes. Animal cells are grown in medium containing growth factors from serum. Growth factors are proteins that promote cell growth and proliferation, and they are essential for the culture of most animal cells. Plating out of a liquid microbial culture on solid media allows the number of colony-forming units to be counted and the density of cells in the culture to be estimated. Serial dilution is often needed to achieve a suitable colony count.

Haemocytometers

Haemocytometers are microscopic grids used to estimate cell numbers in a liquid culture, as shown in Figure 1.17a). When counting cells, a protocol is needed to deal with cells lying across the grid boundaries – often, cells over the right-hand and bottom boundaries of a grid are counted, but those on the top or left-hand boundaries are not. Often, the four corner and central grids in an array of grids are counted and averaged.

Vital staining is required to identify and count viable cells because the stain can distinguish between cells that are alive or dead, as shown in Figure 1.17b).

Hints & tips

Note that the stain in Figure 1.17b) colours the dead cells.

a)

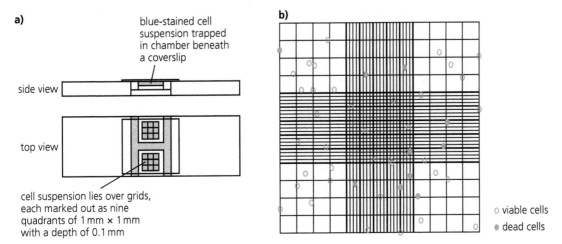

side view

blue-stained cell suspension trapped in chamber beneath a coverslip

top view

cell suspension lies over grids, each marked out as nine quadrants of 1 mm × 1 mm with a depth of 0.1 mm

o viable cells
● dead cells

Figure 1.17 a) Haemocytometer chamber with stained culture trapped against the grids using a coverslip; b) haemocytometer grid with a sample of cells from a culture stained using trypan blue, a vital stain

Check-up 6 ?

1 Explain what is meant by a monoclonal antibody.   1
2 Describe how a reporter enzyme works.   3
3 Describe western blotting.   4
4 Explain why aseptic technique is used when working with cell cultures.   2
5 In Figure 1.17, the volume of the haemocytometer chamber is 0.001 ml. Use this information and the cell number in the whole chamber to calculate:
  a) the number of viable cells
  b) the number of dead cells in 1 ml of the culture.   2

Hints & tips ★

Note that vital stains may stain live cells but others stain dead cells.

Key words

Affinity – the degree to which a substance is attracted and tends to bind to another
Affinity chromatography – a technique used to separate and purify proteins based on a specific binding interaction between an immobilised ligand and its binding partner
Antigen – a specific protein with which antibodies can bind in an immune response
Aseptic technique – procedures in place to prevent contamination, including sterilising equipment and work surfaces
Bright-field microscopy – microscopy commonly used to observe whole organisms, parts of organisms, thin sections of dissected tissue or individual cells
Buffer – a solution used to set and maintain a particular pH
Centrifugation – a process that uses centrifugal forces to separate components of different densities in a mixture
Chromatography – a technique used to separate different substances; it has a stationary phase (for example, paper or gel) that the mobile phase (for example, a solvent) moves through, carrying the substance being examined; different distances are moved by substances of different solubility
Colorimeter – a device used to measure the absorbance of a specific wavelength of light by a solution
Culture media – a nutrient-rich growth medium providing the basic requirements for cell growth (amino acids, glucose, salts, water, as well as specific growth factors for animal cell lines)
Fluorescence – the emission of light of a different wavelength to that which was absorbed
Fluorescence microscopy – microscopy technique that uses specific fluorescent labels to bind to and visualise certain molecules or structures within cells or tissues
Gel electrophoresis – technique used to separate samples of nucleic acid and protein by size; introduced to a gel, they move through it due to an electric current; smaller fragments move further than larger fragments
Growth factors – proteins that promote cell growth and proliferation

⇨

Haemocytometer – microscopic grid used to estimate the total number of cells within a sample (originally used to count blood cells)

Hazard – anything that poses a potential risk or threat to an individual or the environment

Immunoassay – technique used to detect and identify specific proteins; antibodies linked with reporter enzymes, for example, cause a colour change in the presence of a specific antigen

Inoculum – starting material used to grow a culture from, for example a bacterial culture

Isoelectric point (IEP) – the pH at which a soluble protein has no net charge and will precipitate out of solution

Linear dilution series – a series of dilutions that differ by an equal interval, for example 0.1 M, 0.2 M, 0.3 M, and so on

Log dilution series – a series of dilutions that differ by a constant proportion, for example 10^{-1}, 10^{-2}, 10^{-3}, and so on

Monoclonal antibodies – stocks of identical antibodies that are specific to a particular antigen

Native gel electrophoresis – does not contain SDS and does not denature the molecule, so proteins are separated by their shape, size and charge

Pathogenic – causing disease

Primary cell lines – a culture of cells isolated directly from animal or plant tissues; they have a finite lifespan and limited expansion capacity

Reporter enzyme – an enzyme linked to an antibody specific to a protein antigen; they are used in immunoassay techniques

SDS–PAGE – electrophoresis in which the gel contains SDS, which gives all the molecules an equally negative charge and denatures them, separating proteins by size alone

Serum – vitally important as a source of growth factors, hormones, lipids and minerals for the culture of cells

Standard curve – a graph that can be used to determine the concentration of an unknown solution

Supernatant – the liquid that lies above a solid residue or pellet in centrifugation

Turbidity – a measure of the degree to which a fluid loses its transparency due to the presence of suspended particles or cells in suspension

Vital staining – a technique in which a harmless dye is used to stain either living tissue cells or dead cells for microscopical observation to allow a viable cell count to be made

Western blotting – an analytical technique used to identify and locate specific proteins in a sample of tissue homogenate or extract based on their ability to bind to specific antibodies

Exam-style questions

Structured questions

1 The **diagram** below shows a haemocytometer counting chamber containing a sample from a culture of animal cells prepared by an aseptic technique and stained with a vital stain. The depth of the chamber is 0.01 cm.

a) Calculate the density of cells per cm³ in this culture based on the central square. **1**

b) The cells have been stained using a vital stain. Explain what is meant by a vital stain in this example. **1**

c) Describe the components of the culture medium that could ensure the normal growth and proliferation of cells in the culture. **2**

d) The culture was prepared using an aseptic technique. Explain what is meant by an aseptic technique. **2**

2 In a procedure to purify an enzyme, a tissue sample was taken through a number of stages.

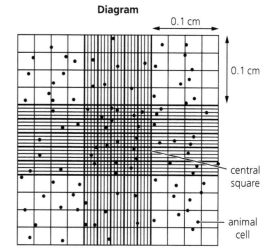

Diagram

The table below describes the purification stages and shows the total mass of protein present and the enzyme activity in the sample following each stage in the purification procedure.

Stage	Description of purification stage	Total protein (mg)	Enzyme activity (units)
1	Liquidise tissue sample	10 000	2 000 000
2	Precipitation by salts	3 000	1 500 000
3	Separation by isoelectric point	500	500 000
4	Separation by affinity chromatography	30	42 000

a)

(i) Calculate the percentage of the protein that had been removed from the liquidised tissue by the end of Stage 4. **1**

(ii) Enzyme purity in a sample can be calculated using the formula below:

$$\text{Enzyme purity} = \frac{\text{(enzyme activity)}}{\text{(total protein)}}$$

Use the formula to calculate the number of times by which enzyme purity had been increased between the liquidised sample and the end of Stage 4. **1**

b) Explain how separation by isoelectric point, as in Stage 3, occurs. **2**

c) In affinity chromatography at Stage 4, a ligand specific to the enzyme being purified was bonded to agarose gel beads packed into a column.
Describe how this method can improve the purity of the enzyme. **2**

Extended response

3 Give an account of laboratory separation techniques used in the separation of amino acids and proteins. **7**

4 Give an account of immunoassay techniques. **7**

Answers are given on pages 147–148.

Key Area 1.2
Proteins

Key points !

1 The **proteome** is the entire set of proteins expressed by a genome. ☐
2 The proteome is larger than the number of genes, particularly in eukaryotes, because more than one protein can be produced from a single gene as a result of **alternative RNA splicing**, in which **introns** are removed from RNA transcripts and **exons** are retained. ☐
3 Not all genes are expressed as proteins in a particular cell type. ☐
4 Genes that do not code for proteins are called **non-coding RNA genes** and include those that are transcribed to produce tRNA, rRNA and RNA molecules that control the expression of other genes. ☐
5 The set of proteins expressed by a given cell type can vary over time and under different conditions. ☐
6 Some factors affecting the proteins expressed by a given cell type are the metabolic activity of the cell, cellular stress, the response to signalling molecules, and diseased versus healthy cells. ☐
7 Eukaryotic cells, because of their size, have a relatively small surface area to volume ratio. ☐
8 The plasma membrane of eukaryotic cells is therefore too small an area to carry out all of the vital functions carried out by membranes. ☐
9 Eukaryotic cells have a system of internal membranes that increase the total area of membrane. ☐
10 The **endoplasmic reticulum (ER)** forms a network of membrane tubules continuous with the nuclear membrane. ☐
11 The **Golgi apparatus** is a series of flattened membrane discs. ☐
12 **Lysosomes** are membrane-bound organelles containing a variety of **hydrolases** that digest proteins, lipids, nucleic acids and carbohydrates. ☐
13 **Vesicles** transport materials between membrane compartments. ☐
14 The components of membranes, **phospholipids** and proteins are synthesised in the ER. Phospholipids are a type of lipid that form the main component of the cell membrane. ☐
15 Lipids are synthesised in the **smooth endoplasmic reticulum (SER)** and inserted into its membrane. ☐
16 **Rough endoplasmic reticulum (RER)** has ribosomes on its cytosolic face, while smooth ER (SER) lacks ribosomes. ☐
17 The synthesis of all proteins begins in cytosolic ribosomes. ☐
18 The synthesis of cytosolic proteins is completed there, and these proteins remain in the cytosol. ☐
19 Transmembrane proteins carry a **signal sequence** that halts translation and directs the ribosome synthesising the protein to dock with the ER, forming RER. ☐
20 A signal sequence is a short stretch of amino acids at one end of a polypeptide that determines the eventual location of a protein in a cell. ☐
21 Translation continues after docking, and the protein is inserted into the membrane of the ER. ☐
22 Once the proteins are in the ER, they are transported by vesicles that bud off from the ER and fuse with the Golgi apparatus. ☐
23 As proteins move through the Golgi apparatus, they undergo **post-translational modification**. ☐
24 Molecules move through the Golgi discs in vesicles that bud off from one disc and fuse to the next one in the stack. ☐
25 Within the Golgi apparatus, enzymes catalyse the addition of various sugars (carbohydrates) in multiple steps to form **glycoproteins**. ☐
26 The addition of carbohydrate groups is the major post-translational modification within the Golgi apparatus. ☐
27 Vesicles that leave the Golgi apparatus take transmembrane proteins to the plasma membrane and lysosomes. ☐

⇨

28 Vesicles move along microtubules to other membranes and fuse with them within the cell. ☐

29 Proteins for secretion are translated in ribosomes on the RER and enter its lumen. ☐

30 Peptide hormones and digestive enzymes are examples of proteins for secretion. ☐

31 The proteins move through the Golgi apparatus and are then packaged into secretory vesicles. ☐

32 Secretory vesicles move to, and fuse with, the plasma membrane, releasing the proteins out of the cell. ☐

33 **Proteolytic cleavage** is another type of post-translational modification. ☐

34 Digestive enzymes are one example of secreted proteins that require proteolytic cleavage of inactive precursors to become active. ☐

35 Proteins are **polymers** of amino acid **monomers**. ☐

36 Amino acids are linked by peptide bonds to form polypeptides. ☐

37 Amino acids have the same basic structure, differing only in the **R group** present. ☐

38 R groups of amino acids vary in size, shape, charge, hydrogen bonding capacity and chemical reactivity. ☐

39 Amino acids are classified according to their R groups: **basic** (positively charged), **acidic** (negatively charged), **polar** or **hydrophobic**. ☐

40 The wide range of functions carried out by proteins results from the diversity of R groups. ☐

41 The **primary structure** is the sequence in which the amino acids are synthesised into the polypeptide. ☐

42 **Hydrogen bonding** along the backbone of the protein strand results in regions of **secondary structure – alpha helices**, parallel or anti-parallel **beta-pleated sheets**, or **turns**. ☐

43 The polypeptide folds into a **tertiary structure** stabilised by interactions between R groups: hydrophobic interactions, **ionic bonds**, **London dispersion forces**, hydrogen bonds and **disulfide bridges**. ☐

44 Disulfide bridges are covalent bonds between R groups containing sulfur. ☐

45 **Quaternary structure** exists in proteins with two or more connected polypeptide subunits. ☐

46 The quaternary structure describes the spatial arrangement of the subunits. ☐

47 A **prosthetic group** is a non-protein unit tightly bound to a protein and necessary for its function. ☐

48 The ability of haemoglobin to bind with oxygen is dependent on the prosthetic haem group. ☐

49 Interactions of R groups can be influenced by temperature – increasing temperature disrupts the interactions that hold the protein in shape; the protein begins to unfold, eventually becoming denatured. ☐

50 The charges on acidic and basic R groups are affected by pH – as pH increases or decreases from the optimum, the normal ionic interactions between charged groups are lost, which gradually changes the **conformation** of the protein until it becomes denatured. ☐

51 A **ligand** is a substance that can bind to protein R groups that are not involved in protein folding. ☐

52 Binding sites have a complementary shape and chemistry to the ligand. ☐

53 As a ligand binds to a protein-binding site, the protein conformation changes. This change causes a functional change in the protein. ☐

54 Allosteric interactions occur between spatially distinct sites. The binding of a substrate molecule to one active site of an **allosteric enzyme** increases the affinity of the other active sites for binding of subsequent substrate molecules. This is of biological importance because the activity of allosteric enzymes can vary greatly with small changes in substrate concentration. ☐

55 Many allosteric proteins consist of multiple subunits (quaternary structure). ☐

56 Allosteric proteins with multiple subunits show **co-operativity** in binding, in which changes in binding at one subunit alter the affinity of the remaining subunits. ☐

57 Allosteric enzymes contain a second type of site, called an allosteric site. ☐

58 **Modulators** regulate the activity of the enzyme when they bind to the allosteric site. ☐

59 Following binding of a modulator, the conformation of the enzyme changes, which alters the affinity of the active site for the substrate. ☐

60 Positive modulators increase the enzyme's affinity for the substrate; negative modulators reduce the enzyme's affinity. ☐

61 The binding and release of oxygen in haemoglobin shows co-operativity – changes in binding of oxygen at one subunit alter the affinity of the remaining subunits for oxygen. ☐

62 Temperature and pH have an influence on the binding of oxygen to haemoglobin. ☐

63 A decrease in pH or an increase in temperature lowers the affinity of haemoglobin for oxygen, so the binding of oxygen is reduced. Reduced pH and increased temperature in actively respiring tissue will reduce the binding of oxygen to haemoglobin promoting increased oxygen delivery to tissue. ☐

64 The addition or removal of phosphate can cause reversible conformational change in proteins. ☐

65 Phosphorylation is a common form of post-translational modification. ☐

66 **Protein kinases** catalyse the transfer of a phosphate group to other proteins. ☐

67 The terminal phosphate of ATP is transferred to specific R groups. ☐

68 **Protein phosphatases** catalyse the removal of phosphate from a protein. ☐

69 Phosphorylation brings about conformational changes, which can affect a protein's activity. ☐

70 The activity of many cellular proteins, such as enzymes and receptors, is regulated by phosphorylation. ☐

71 Some proteins are activated by phosphorylation while others are inhibited. ☐

72 Adding a phosphate group adds negative charges. ☐

73 Ionic interactions in the unphosphorylated protein can be disrupted and new ones created. ☐

The proteome

The proteome is the entire set of proteins expressed by a genome. The proteome is larger than the number of genes, particularly in eukaryotes, because more than one protein can be expressed from a single gene due to alternative RNA splicing, as shown in Figure 1.18. Different exons (expressons) can be expressed from within one gene, or different introns (intragenic regions) may be retained depending on the splicing process.

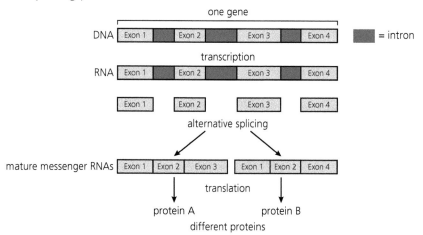

Figure 1.18 The role of alternative splicing in the production of different proteins from a single gene

Not all genes are expressed as proteins in a particular cell, and the set of proteins expressed by a given cell type can vary over time and under different conditions. Some factors affecting the set of proteins expressed by a given cell type are:

- the metabolic activity of the cell, which changes with age, senescence, dormancy state, and so on
- its state of cellular stress depending on extremes of temperature, pH, exposure to toxins, mechanical damage, and so on
- its response to signalling molecules such as hormones and, in the case of lymphocytes, the antigens to which it is exposed
- its state of health or disease and/or during apoptosis.

Hints & tips ⭐

Make sure you know how RNA splicing can produce alternative proteins from the expression of the same gene — not much more knowledge than you needed at Higher here.

Types of gene

Some genes code for proteins. Those that do not are called non-coding RNA genes and include those that are transcribed to produce tRNA, rRNA and various other RNA molecules that control the expression of protein-encoding genes.

Check-up 7 ❓

1 Describe what is meant by the proteome.   1
2 Explain why the proteome is larger than the genome in many eukaryotic organisms.   1
3 Describe alternative RNA splicing.   2
4 List **three** factors that can affect the set of proteins expressed by a cell type.   3

The synthesis and transport of proteins

Intracellular membranes

Because of their relatively large size, eukaryotic cells have a relatively small surface area to volume ratio. The surface area of their plasma membrane is too small to carry out all the vital functions that rely on membranes and the specialised proteins associated with them.

To increase the total membrane area they have, they have an internal system of specialised membranes called the endoplasmic reticulum (ER), which forms a network of membrane tubules continuous with the nuclear membrane as shown in Figure 1.19.

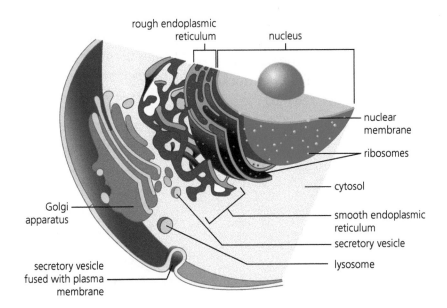

Figure 1.19 Membrane system of a eukaryotic cell

The ER is either rough (RER) or smooth (SER). RER has docked ribosomes on its cytosolic face, while SER lacks ribosomes. The Golgi apparatus is a series of flattened membrane discs related to the ER and has associated vesicles that transport materials between membrane compartments. Lysosomes are formed from specialised Golgi vesicles. Lysosomes are membrane-bound organelles containing a variety of hydrolases that can digest proteins, lipids, nucleic acids and carbohydrates.

Synthesis of membrane components

Phospholipids and proteins are the main membrane components, as shown in Figure 1.20. Phospholipid molecules have a hydrophilic head and hydrophobic tails, and are formed into a bilayer.

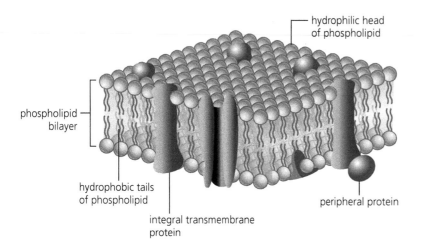

- hydrophilic head of phospholipid
- phospholipid bilayer
- hydrophobic tails of phospholipid
- integral transmembrane protein
- peripheral protein

Figure 1.20 Structure of the cell membrane – note that the protein components of the membrane are either integral and permanently attached or peripheral and only temporarily attached either to exposed integral protein or lipids

Lipids are synthesised in the SER and inserted into its membrane. The synthesis of all proteins begins in cytosolic ribosomes and the synthesis of cytosolic proteins is completed there. These proteins remain in the cytosol (cytoplasm) of the cell where they carry out their specific functions. The synthesis of transmembrane proteins begins in cytosolic ribosomes but is completed when the relevant cytosolic ribosomes dock with the ER to become part of the RER.

Transmembrane proteins are integral to the membrane and are permanently attached there. Not all integral proteins are completely transmembrane. Peripheral proteins are not embedded and form weak bonds on the surfaces of the membrane, either with the phospholipid heads or with the exposed parts of integral proteins.

Transmembrane proteins carry a signal sequence. A signal sequence is a short stretch of 16–30 amino acids at one end of the polypeptide that will determine the eventual location of that protein in a cell. The signal sequence is at the N terminus of the protein and so is synthesised first. If present, the signal sequence halts translation and directs the ribosome synthesising the protein to dock with the ER, forming RER. After docking, the signal sequence is removed and the protein is inserted into the membrane of the ER, as shown in Figure 1.21.

> **Key links** 👍
>
> There is more about transmembrane proteins in Key Area 1.3.

> **Key links** 👍
>
> See Figure 1.25 for a definition of the N terminus of a protein.

> **Check-up 8** ❓
>
> 1 Describe the various different membranes inside a cell. **4**
> 2 Name the chemical components of a cell membrane and their synthesis. **3**

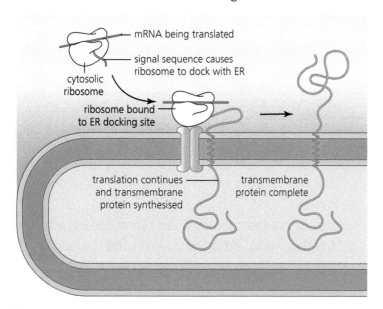

- mRNA being translated
- signal sequence causes ribosome to dock with ER
- cytosolic ribosome
- ribosome bound to ER docking site
- translation continues and transmembrane protein synthesised
- transmembrane protein complete

Figure 1.21 Stages in the translation of a transmembrane protein

Once the proteins are in the ER membrane, they are transported in the membranes of vesicles that bud off from the ER and fuse with the Golgi apparatus, as shown in Figure 1.23.

Post-translational modification

As proteins move through the Golgi apparatus, they undergo post-translational modification. The addition of carbohydrate groups is the major modification involved. Enzymes catalyse the addition of various sugars in multiple steps to form the added carbohydrates and glycoproteins are produced. Molecules move through the Golgi discs in vesicles that bud off from one disc and fuse to the next one in the stack. The glycoprotein-containing vesicles can be recruited into other membranes, including the plasma membrane, as shown in Figure 1.23.

The secretory pathway

Proteins for secretion, such as peptide hormones and digestive enzymes, or those that will become lysosome hydrolases, are translated in ribosomes docked on the RER but enter its lumen rather than becoming integrated with the lipid components, as shown in Figure 1.22.

These proteins move through the Golgi apparatus and are then packaged inside secretory vesicles. Some vesicles that leave the Golgi apparatus take proteins to the plasma membrane for secretion from the cell, while some develop into lysosomes that are retained within the cytosol, as shown in Figure 1.23.

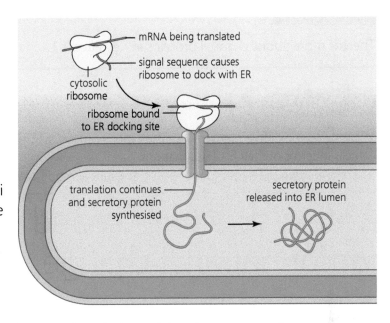

Figure 1.22 Stages in the translation of a secretory protein

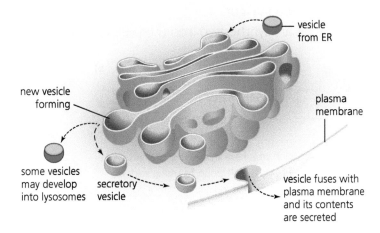

Figure 1.23 Role of the Golgi apparatus in secretion and in the formation of lysosomes

Many proteins which are to be secreted are synthesised as inactive precursors and require proteolytic cleavage to produce active proteins. Proteolytic cleavage is another type of post-translational modification. Digestive enzymes are one example of secreted proteins that require proteolytic cleavage of their precursors to become active. If digestive enzymes were synthesised in active form, they could digest the tissues in which they were synthesised.

Protein	Synthesis notes	Examples
Intracellular protein	Translated completely by cytosolic ribosomes and retained within the cell	Enzymes that control glycolysis
Transmembrane protein	Partially translated by cytosolic ribosomes; signal sequence causes ribosome to dock with ER; translation completed and protein inserted into ER membrane; carried by vesicles to destination	Channel proteins to permit the transport of specific substances across the cell membrane
Extracellular protein for secretion	Translated by docked ER ribosome; enter lumen of ER and move through into Golgi apparatus and into vesicles for secretion or into vesicles that become lysosomes	Peptide hormones such as insulin, enzymes such as salivary amylase, and hydrolases that enter lysosomes

Key links

There is more about channel proteins in Key Area 1.3.

Check-up 9

1 Describe the differences between the various protein synthesised in cells in terms of their final destinations. **4**
2 Describe **two** different types of post-translational modification. **2**

Protein structure, ligand binding and conformational change

Protein structure

Proteins are polymers of amino acid monomers. Amino acids, as shown in Figure 1.24a), link by peptide bonds to form polypeptides, as shown in Figure 1.24b). The primary structure of a protein is the sequence in which the amino acids are synthesised into the polypeptide.

a)

amine group — hydrogen atom

the R-group is different in different amino acids — carboxylic acid group

b)

amino acid   amino acid

condensation → H_2O

peptide bond

Figure 1.24 a) Structure of an amino acid – all amino acids have an amino group and a carboxylic acid group as shown, but differ depending on their R group; b) amino acids are linked by the removal of a water molecule, making a peptide bond and forming a polypeptide

Amino acids differ from each other in their R groups as shown in the table below. Some are described as non-polar – these are hydrophobic and tend to be insoluble in water. The others are hydrophilic and more soluble in water. Those with positively charged side groups are basic, while those with negatively charged side groups are acidic. Those hydrophilic amino acids with side groups that have a very small charge (which cancels out to zero net charge) are polar. Interactions between R groups are very important in determining the overall shapes of proteins, and the R groups can influence the position of protein molecules in cells.

R group classification	R group description	Affinity with water	Notes	Example
Non-polar	Mostly carbon and hydrogen atoms	Hydrophobic – repelled by water	Attracted to each other	Alanine (ala)
Basic (positive charge)	Additional amino group gives net positive charge	Hydrophilic – attracted to water	Can form ionic bonds	Lysine (lys)
Acidic (negative charge)	Additional acidic group gives net negative change	Hydrophilic – attracted to water	Can form ionic bonds	Glutamic acid (glu)
Polar (neutral)	Uncharged because of a balance of tiny charges	Hydrophilic – attracted to water	Can form hydrogen bonds	Serine (ser)

Primary structure

The sequence or order of specific amino acids along the polypeptide chain, as shown in Figure 1.25, is the primary structure. This is the order that is determined by the base sequence in mature mRNA.

Secondary structure

Hydrogen bonds are formed when weak positive charges on hydrogen atoms are attracted to weak negative charges on oxygen or nitrogen atoms. Hydrogen bonding occurs along the backbone of the protein strand and results in regions of secondary structure – alpha helices, parallel or anti-parallel beta-pleated sheets, as shown in Figure 1.26a), or turns. Turns usually join different secondary structures together or allow changes of direction in the polypeptide, causing folding to create compact molecules. Many proteins have all of these secondary features within their molecule, as shown in Figure 1.26b).

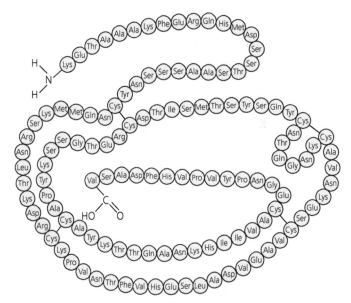

Figure 1.25 Primary structure of a protein – the sequence of amino acids in the polypeptide chain; note that the end of the protein, which has an amine group (NH_2), is called the N terminus

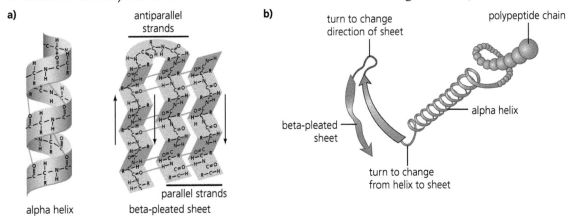

Figure 1.26 a) Hydrogen bonding in helices and beta-pleated sheets – note the arrowed directional difference between the parallel and anti-parallel sheets; b) ribbon diagram to show the different secondary structures in a protein – the amino acids in the ribbons are held in the secondary structure by hydrogen bonds between R groups as shown in a)

Check-up 10 ?

1 Name the **four** different types of R group and describe the characteristics of each.   **4**
2 Describe the primary and secondary level of protein structure.   **3**

Tertiary structure

The polypeptide folds into a tertiary structure, which gives the molecule an overall shape in space. This conformation is stabilised by interactions between R groups. These interactions, shown in Figure 1.27, include:

- hydrophobic interactions, which occur between non-polar side groups that clump together to avoid water and tend to locate on the inside of the molecular structure
- ionic bonds, which occur between R groups with opposite charges
- disulfide bridges, which are covalent bonds between the small numbers of amino acids with R groups containing sulfur
- London dispersion forces, which are weak, temporary attractions between atoms caused by localised movements of their electrons
- hydrogen bonds, which are attractions between hydrogen atoms and either oxygen or nitrogen atoms caused by tiny differences in their charges.

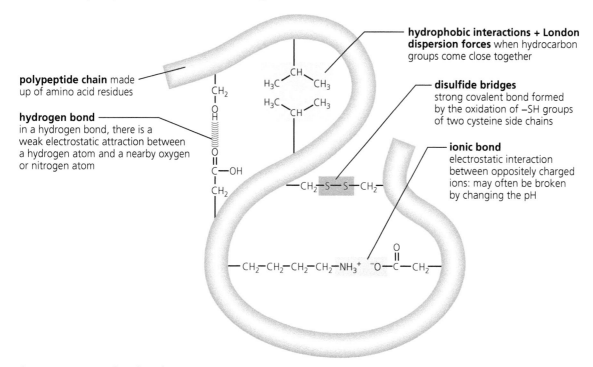

Figure 1.27 Various bonds and interactions between R groups that can hold polypeptide secondary structures into a tertiary structure

Interactions of R groups can be influenced by temperature and pH. Increasing temperature disrupts the interactions that hold the protein in shape and the protein begins to unfold, eventually becoming denatured. The charges on acidic and basic R groups are affected by pH. As pH increases or decreases from the optimum, the normal ionic interactions between charged groups are lost, which gradually changes the conformation of the protein until it becomes denatured.

Quaternary structure

Quaternary structure exists in proteins with two or more connected polypeptide subunits. The subunits are linked to each other by bonds or interactions between certain R groups of the different subunits. Quaternary structure describes the spatial arrangement of these subunits.

A prosthetic group is a non-protein unit tightly bound to a protein and necessary for its specific function. The ability of haemoglobin to bind with oxygen depends on the non-protein haem group, for example. The following table summarises the levels of protein structure.

Level of structure	Definition	Bonding	Diagram
Primary	Sequence of amino acids	Peptide bonds between amino acids to form a polypeptide	
Secondary	Folding of amino acid chain into helices, beta-pleated sheets and turns	Hydrogen bonds between R groups along the polypeptide	
Tertiary	Further folding of secondary shapes into an overall three-dimensional shape	Various bonds and interactions including disulfide bridges, ionic bonds, hydrogen bonds, London dispersion forces and hydrophobic interactions	
Quaternary	Complex shape produced by combining two or more connected polypeptide subunits	Further bonding between R groups on the different polypeptides that make up the subunits	

Ligands

A ligand is a substance that can bind to a protein. R groups not involved in protein folding can allow binding to these other molecules. Binding sites have complementary shapes and chemistry to the ligand. As a ligand binds to a binding site, the conformation of the protein changes. This change in conformation causes a functional change in the protein.

Allosteric interactions occur between spatially distinct sites on a protein's surface. The binding of a substrate molecule to one active site of an allosteric enzyme increases the affinity of the other active sites for binding of subsequent substrate molecules. This is of biological importance because the activity of allosteric enzymes can vary greatly with small changes in substrate concentration. Many allosteric proteins consist of multiple subunits (they have quaternary structure). Allosteric proteins with multiple subunits show co-operativity in binding, in which changes in binding at one subunit alters the affinity of the remaining subunits.

Haemoglobin and oxygen

The binding and release of oxygen in haemoglobin shows co-operativity. This term refers to changes in the affinity of the remaining subunits for oxygen following the binding or release of oxygen from the first subunit. A decrease in pH or an increase in temperature lowers the affinity of haemoglobin for oxygen, so the binding of oxygen is reduced. The reduced pH and increased temperature in active, respiring muscle tissue will reduce the affinity of oxygen for haemoglobin, making it easier for oxygen to be released to muscle tissues, as shown in Figure 1.28.

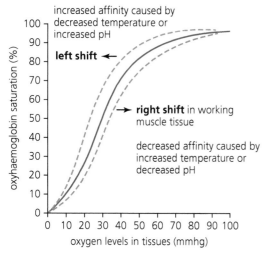

Figure 1.28 Haemoglobin dissociation curve and the effects of increased temperature and increased acidity (decreased pH) which occurs during the increased respiration in working muscle tissue

---

Check-up 11 ?

1 Name **five** types of bond that can hold the tertiary structure of protein in shape. **5**
2 Describe the quaternary structure of a protein. **2**
3 Explain what is meant by co-operativity between subunits of a protein. **2**

Allosteric sites and enzymes

Allosteric enzymes contain a second type of site other than the active site, called an allosteric site. Modulators regulate the activity of the enzyme when they bind to the allosteric site. Following binding of a modulator, the conformation of the enzyme changes and this alters the affinity of the active site for the substrate. Negative modulators (inhibitors) reduce the enzyme's affinity for the substrate, while positive modulators (activators) increase the enzyme's affinity, as shown in Figure 1.29.

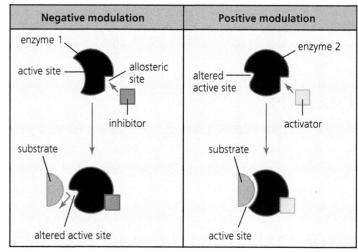

Figure 1.29 Negative and positive modulation of enzyme action – negative modulators act as inhibitors and positive modulators are activators

Phosphorylation

The addition or removal of phosphate from particular R groups can be used to cause reversible conformational changes in proteins. This is another form of post-translational modification. Protein kinases catalyse the transfer of a phosphate group to other proteins. The terminal phosphate of ATP is transferred to specific R groups. Protein phosphatases catalyse the reverse reaction.

Phosphorylation brings about conformation changes that can affect a protein's activity. The activity of many cellular proteins, such as enzymes and receptors, is regulated in this way, as shown in Figure 1.30.

Some proteins are activated by phosphorylation while others are inhibited. Adding a phosphate group adds a negative charge, which can disrupt ionic interactions in the unphosphorylated protein and new ones can be created.

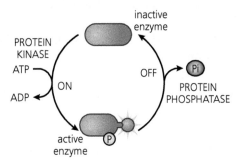

Figure 1.30 Conformational changes brought about by kinases and phosphatases can be used to control an enzyme's activity

Check-up 12 ?

1	Describe co-operativity in haemoglobin.	2
2	Describe allosteric sites and how modulators affect enzyme action.	3
3	Explain how phosphorylation can affect the functioning of proteins.	2

Key words

Allosteric enzymes – enzymes that change conformation in response to a modulator

Alpha helix – polypeptide chain coiled into a helix with hydrogen bonding occurring to maintain the arrangement

Alternative RNA splicing – removal of non-coding introns from a primary mRNA transcript to leave only the coding exons; several different mature transcripts can be produced from a single primary transcript

Beta-pleated sheets – polypeptide chain arranged in rows with the chain folding in parallel or anti-parallel arrangements

Conformation – structural arrangement of the polypeptide chains within a protein; it can be altered by factors such as pH and the binding of ligands and modulators

Co-operativity – changes in binding of a target molecule to one subunit of a multiunit polypeptide changes the affinity of the other subunits for the target molecule

Disulfide bridge – a strong covalent bond that stabilises the tertiary and quaternary structures of many proteins

Endoplasmic reticulum (ER) – a network of membrane tubules within the cytoplasm of a eukaryotic cell, continuous with the nuclear membrane

Exon – section of RNA that is usually retained during splicing

Glycoprotein – a protein with a carbohydrate added by post-translational modification

Golgi apparatus – a series of flattened membrane discs that packages proteins into membrane-bound vesicles inside the cell before the vesicles are sent to their destination

Hydrogen bonds – attractions between polar molecules in which hydrogen is bound to a larger atom, such as oxygen or nitrogen

Hydrolases – a class of enzyme that use water to break chemical bonds

Intron – a section of RNA usually removed during splicing

Ionic bonds – a type of chemical bonding that involves the electrostatic attraction between oppositely charged ions

Ligand – a substance that can bind to a protein; the protein has a shape complementary to the ligand to allow binding

London dispersion force – a temporary, weak attraction between atoms and molecules

Lysosome – a modified Golgi vesicle containing hydrolytic enzymes

Modulators – these bind to a secondary site on an enzyme to alter its conformation; positive modulators activate enzymes and negative modulators deactivate them

Monomer – a molecule that can bind chemically to other monomers to form a polymer

Non-coding RNA gene – a gene that codes for RNAs other than messenger RNA, so do not encode protein

Phospholipid – component of cell membranes

Polymer – a macromolecule composed of many repeated subunits (monomers)

Post-translational modification – addition of different chemical groups to, or modification of, a protein to allow a particular function

Prosthetic group – a non-protein unit tightly bound to a protein and necessary for its function

Protein kinases – catalyse the transfer of a phosphate group from a donor molecule (usually ATP) to an acceptor

Protein phosphatases – an enzyme that removes a phosphate group from its substrate

Protein structure – different levels of arrangement of polypeptides within a protein:
- **primary structure** – sequence in which amino acids are found within a protein
- **secondary structure** – hydrogen bonding occurring within a polypeptide forming alpha helices or beta-pleated sheets
- **tertiary structure** – bonding of many types occurring between the R groups of amino acids within a protein
- **quaternary structure** – the arrangement of multiple folded polypeptides connected together

Proteolytic cleavage – a major form of post-translational modification; it occurs when a protease cleaves one or more bonds in a target protein to activate, inhibit or destroy the protein's activity

Proteome – the entire set of proteins expressed by a genome; it is much larger than the genome

R groups – side groups that allow different bonding between amino acids and give them their wide range of functions:
- **basic R group** – contains an amine functional group and produces a basic solution because the extra amine group is not neutralised by the acidic group
- **acidic R group** – contains an acidic functional group and produces an acidic solution because the extra acid group is not neutralised by the amine group
- **polar R group** – group that prefers to exist in a watery environment
- **hydrophobic R group** – composed mostly of carbon and hydrogen, and tend to be repelled from water

Rough endoplasmic reticulum (RER) – organelle made up of membranes with ribosomes attached

Signal sequence – a short stretch of amino acids at one end of the polypeptide that determines its eventual location in a cell

Smooth endoplasmic reticulum (SER) – a membranous organelle found in most eukaryotic cells; its main functions are the synthesis of lipids and steroid hormones

Turns – secondary structure that reverses the direction of a polypeptide chain

Vesicles – small membrane-bound compartments filled with liquid

Exam-style questions

Structured questions

1 The diagram below shows the structure of Golgi apparatus from a cell which secretes enzymes into saliva.

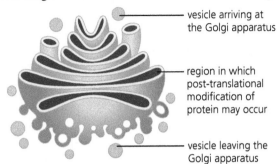

vesicle arriving at
the Golgi apparatus

region in which
post-translational
modification of
protein may occur

vesicle leaving the
Golgi apparatus

   a) Describe the location(s) of translation of the enzymes that are to be secreted from the cell.   **2**
   b) Identify the part of this cell from which the vesicle arriving at the Golgi apparatus has come.   **1**
   c) Describe **one** example of a post-translation modification that might occur within the
      Golgi apparatus.   **1**
   d) Describe how enzymes within vesicles leaving the Golgi apparatus are secreted.   **1**
   e) Give one post-translational modification to an enzyme that happens following secretion.   **1**

2 The diagram opposite shows the structure of a haemoglobin molecule.

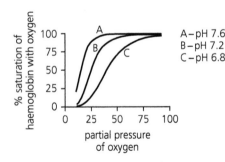

haem group
iron atom

region X

   a) Identify region X, which is part of the secondary structure of a
      subunit of haemoglobin.   **1**
   b) With reference to the haemoglobin molecule, describe what is
      meant by the quaternary structure of a protein.   **2**
   c) Give the term used to describe non-protein groups, such as
      haem, which are tightly bound to proteins and are essential for
      their function.   **1**

3 The graph below shows how the binding of oxygen to human haemoglobin is affected by the partial
   pressure of oxygen and the pH in its surroundings.

   a) Explain how the principle of positive co-operativity can
      account for the general shapes of graphs A, B and C.   **3**
   b) Graph C represents the percentage saturation of
      haemoglobin with oxygen in muscle tissue.
      (i) Name **two** substances produced by rapidly respiring
          muscle cells that would reduce its pH.   **2**
      (ii) Explain why the change in percentage saturation
           of haemoglobin with oxygen at a lower pH is an
           advantage in muscle tissue.   **2**
      (iii) Apart from lowering the pH, give one other factor
            that could shift the curve of haemoglobin to the
            right.   **1**

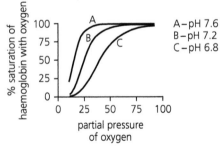

A – pH 7.6
B – pH 7.2
C – pH 6.8

% saturation of
haemoglobin with oxygen

partial pressure
of oxygen

Extended response

4 Describe how phosphate groups can affect the structure and function of proteins.   **5**
5 Give an account of the different classes of amino acids and explain their roles in the tertiary
   structure of proteins.   **8**

Answers are given on page 149.

Key Area 1.3
Membrane proteins

Key points !

1 The **fluid mosaic model** describes the structure of the plasma membrane as a mosaic of components – including phospholipids, cholesterol, proteins and carbohydrates – that gives the membrane a fluid character. ☐

2 Regions of hydrophobic R groups allow extensive hydrophobic interactions that hold **integral membrane proteins** within the phospholipid bilayer. ☐

3 Some integral membrane proteins are **transmembrane proteins**, which span across the membrane. ☐

4 **Peripheral membrane proteins** have hydrophilic R groups on their surface and are bound to the surface of membranes, mainly by ionic and hydrogen bond interactions. ☐

5 Many peripheral membrane proteins interact with the surfaces of integral membrane proteins. ☐

6 The phospholipid bilayer is a barrier to ions and most uncharged polar molecules. ☐

7 Some small molecules, such as oxygen and carbon dioxide, pass through the bilayer by simple diffusion. ☐

8 **Facilitated diffusion** is the passive transport of substances across the membrane through specific transmembrane proteins. ☐

9 To perform specialised functions, different cell types have different channel and **transporter proteins**. ☐

10 Most channel proteins in animal and plant cells are highly selective. ☐

11 Channels are multi-subunit proteins with the subunits arranged to form water-filled pores that extend across the membrane. ☐

12 Some channel proteins are **gated** and change conformation to allow or prevent diffusion. ☐

13 **Ligand-gated channels** are controlled by the binding of signal molecules; **voltage-gated channels** are controlled by changes in ion concentration. ☐

14 Transporter proteins bind to the specific solute to be transported and undergo a conformational change to transfer the substance across the membrane. ☐

15 Transporters alternate between two conformations so that the binding site for a solute is sequentially exposed on one side of the bilayer, then the other. ☐

16 Active transport uses pump proteins that transfer substances across the membrane against their concentration gradient. ☐

17 Pumps that mediate active transport are transporter proteins coupled to a source of metabolic energy required for active transport. ☐

18 Some active transport proteins are ATPases which hydrolyse ATP directly to provide the energy for the conformational change required to move substances across the membrane. ☐

19 For a solute carrying a net charge, the concentration gradient and the electrical potential difference combine to form the **electrochemical gradient** that determines the transport of the solute. ☐

20 A **membrane potential** (an electrical potential difference) is created when there is a difference in electrical charge on the two sides of the membrane. ☐

21 Ion pumps, such as the **sodium–potassium pump**, use energy from the hydrolysis of ATP to establish and maintain ion gradients. ☐

22 For each ATP hydrolysed, three sodium ions are transported out of the cell and two potassium ions are transported into the cell, establishing both concentration and electrical gradients. ☐

23 The sodium–potassium pump transports ions against a steep concentration gradient using energy directly from ATP hydrolysis to actively transport sodium ions out of the cell and potassium ions into the cell. ☐

⇨

24 The sodium–potassium pump has a high affinity for sodium ions inside the cell where binding occurs; phosphorylation by ATP causes conformational changes, resulting in a decrease in affinity for sodium ions, which are released outside of the cell. The pump has a high affinity for potassium ions outside the cell where binding occurs; following dephosphorylation and conformational changes, potassium ions are released into the cell and the affinity returns to the start. □

25 The sodium–potassium pump in intestinal epithelial cells generates a sodium ion gradient across their plasma membrane and drives the active transport of glucose. □

26 The sodium–potassium pump is found in most animal cells, accounting for a high proportion of the basal metabolic rate in many organisms. □

27 The **glucose symport** transports sodium ions and glucose at the same time, and in the same direction: sodium ions enter the cell down their concentration gradient while the simultaneous transport of glucose pumps glucose into the cell against its concentration gradient. □

Movement of molecules across membranes

The structure of the plasma membrane is usually described using the fluid mosaic model. The membrane is formed by a phospholipid bilayer, which gives it a fluid character and provides flexibility, and proteins, which form a mosaic pattern within the phospholipid bilayer. The highly hydrophobic parts of phospholipid molecules make up the central region of the bilayer. The surfaces of membranes are hydrophilic, and are attracted to the watery cytosol of cells and to the fluid on the extracellular side of the membrane, as shown in Figure 1.31.

The phospholipid bilayer acts as a barrier to ions and most uncharged polar molecules. Some small, non-polar molecules, such as oxygen and carbon dioxide, can pass through the bilayer by simple diffusion, however.

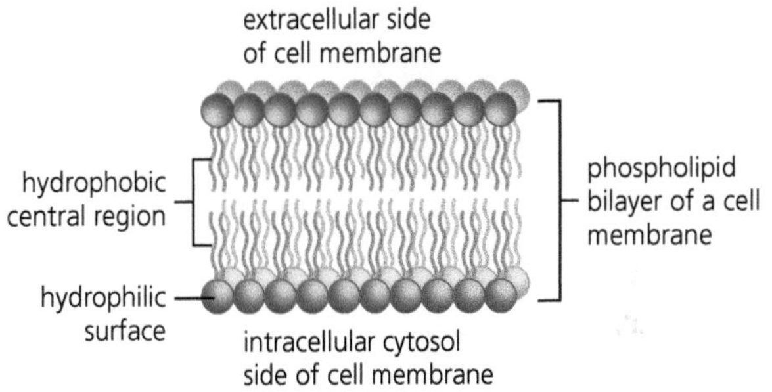

Key links

There is more about hydrophobic molecules in Key Area 1.2.

Figure 1.31 The phospholipid bilayer of a membrane showing its hydrophobic central region and hydrophilic surfaces

Integral proteins

Regions of hydrophobic R groups on their amino acids allow integral proteins to be held permanently within the bilayer because of strong hydrophobic interactions with it. Some integral membrane proteins are transmembrane proteins, which are completely embedded in the membrane and exposed at both ends, as shown in Figure 1.32.

Facilitated diffusion is the passive transport of substances across the membrane through specific transmembrane proteins that have channels within their structure.

Peripheral proteins

Peripheral membrane proteins have hydrophilic R groups in amino acids on their surface and are bound temporarily to the surface of membranes, mainly by ionic and hydrogen bond interactions. Many peripheral membrane proteins interact with the surfaces of integral membrane proteins.

Check-up 13

1 Name the chemical components of a cell membrane.   3
2 Explain how different proteins bind to the membranes of cells.   3

Channels and transporters

Specific transmembrane proteins can act as channels and others as transporters to control ion concentrations and concentration gradients. To perform specialised functions, different cell types have different channel and transporter proteins. Most channel proteins in animal and plant cells are highly selective and promote the passage of specific molecules by facilitated diffusion. Channels are multi-subunit proteins with the subunits arranged to form water-filled pores that extend across the membrane.

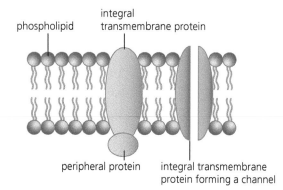

Figure 1.32 Membrane showing integral and peripheral proteins associated with the phospholipid bilayer

Figure 1.33 shows examples of the transport of various substances through the cell membrane.

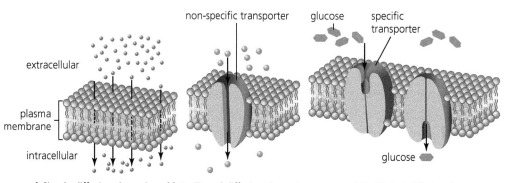

a) Simple diffusion through the lipid bilayer

b) Facilitated diffusion through a non-specific transporter

c) Facilitated diffusion through a specific transporter

Figure 1.33 Various movements across the cell membrane through lipid and channel proteins – these movements depend of the size and solubility of the substances as well as any specific transporter proteins involved

Check-up 14 ?

1 Describe the roles of membrane proteins in facilitated diffusion. **2**

Transporters

Transporter proteins bind to the specific substance to be transported and undergo a conformational change to transfer the solute across the membrane. Active transport uses energy to transfer substances across the membrane against their concentration gradient. Pumps that mediate active transport are transporter proteins coupled to an energy source. Some active transport proteins are ATPases, which hydrolyse ATP directly to provide the energy for the conformational change required to move substances across the membrane, as shown in Figure 1.34.

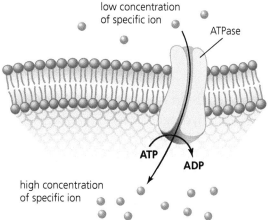

Figure 1.34 A specific ion is transported against its concentration gradient through a transporter called ATPase, which uses energy from ATP directly

Key links 👍

There is more about conformational change in Key Area 1.2.

Gated channels

Some channel proteins are gated and change conformation to allow or prevent diffusion, for example sodium channels and potassium channels. Ligand-gated channels can be controlled by signal molecules that act as ligands, binding specifically to them; voltage-gated channels rely on changes in ion concentrations, as shown in Figure 1.35.

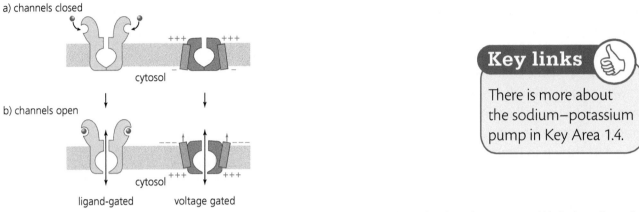

Figure 1.35 Membrane showing a) a ligand-gated and a voltage-gated channel in their closed conformation, and b) the ligand-gated channel in its open conformation after the binding of ligands and the a voltage-gated channel in its open conformation following a change in ion concentration across the membrane

> **Key links** 👍
>
> There is more about the sodium–potassium pump in Key Area 1.4.

> **Check-up 15** ❓
>
> 1 Explain the difference between a ligand-gated channel and a voltage-gated channel in a cell membrane. **4**
> 2 Describe how an ATPase transporter protein works. **3**

Ion transport pumps and generation of ion gradients

For a solute carrying a net charge, the concentration gradient and the electrical potential difference combine to form the electrochemical (ion) gradient that determines the direction of transport of the solute. A membrane potential is created when there is an electrical potential difference between the two sides of a membrane.

Ion pumps, such as the sodium–potassium pump, use energy from the hydrolysis of ATP to establish and maintain ion gradients. The sodium–potassium pump transports both sodium and potassium ions against a steep concentration gradient using energy directly from ATP hydrolysis. It actively transports sodium ions out of the cell and potassium ions into the cell, as shown in Figure 1.36.

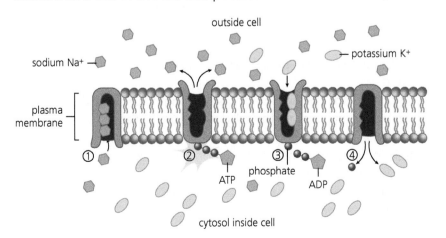

Figure 1.36 ① the pump transporter has a high affinity for sodium ions; it binds three sodium ions on the cytosol side of the membrane; ② it is phosphorylated by splitting ATP and changes its conformation, releasing the three sodium ions to the outside of the cell; ③ the changed conformation gives the protein high affinity for potassium outside the cell, and two potassium ions bind; ④ the protein becomes dephosphorylated and reverts to its original conformation, releasing the two potassium ions to the inside of the cell

Glucose transport through intestinal cells

In intestinal epithelial cells, the sodium–potassium pump generates a sodium ion gradient across the plasma membrane so that there is always a low concentration of sodium inside the cell. Sodium ions can then enter the cell down their concentration gradient through a glucose symport channel; the energy from the flow allows the simultaneous transport of glucose into the cell against its concentration gradient, from where it can diffuse into the blood as shown in Figure 1.37.

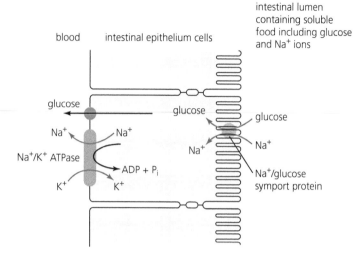

Figure 1.37 Action of glucose symport and sodium–potassium ATPase in the movement of glucose and ions in the intestine; note that sodium–potassium ATPase maintains the concentration gradient required for the symport to function

Check-up 16

1 Describe how the sodium–potassium pump creates an ion concentration gradient across a cell membrane. **4**
2 Explain how the glucose symport allows the movement of glucose molecules against their concentration gradient. **3**

Hints & tips

The glucose symport requires a concentration gradient of sodium ions from outside to inside – anything that reduces that gradient reduces the activity of the symport.

Key words

Electrochemical gradient – a gradient of electrochemical potential, usually for an ion that can move across a membrane, consisting of the difference in solute concentration and the difference in charge across a membrane

Facilitated diffusion – the passive transport of substances across the membrane through specific transmembrane proteins

Fluid mosaic model – a model that describes the structure of the plasma membrane as a mosaic of components including a phospholipid bilayer, which gives the membrane a fluid character, and cholesterol, proteins and carbohydrates

Gated channels – channel-forming proteins controlled by signalling molecules or ion concentration

Glucose symport – an integral membrane protein involved in transport of glucose and sodium ions across the cell membrane at the same time and in the same direction

Integral membrane proteins – also called intrinsic proteins; they have one or more segments that are embedded in the phospholipid bilayer

Ligand-gated channels – transmembrane protein channels controlled by the binding of signal molecules

Membrane potential – an electrical gradient that forces ions to move passively in one direction; positive ions are attracted by the 'negative' side of the membrane and negative ions by the 'positive' one

Peripheral membrane proteins – membrane proteins that adhere only temporarily to either side of a membrane with which they are associated

Sodium–potassium ATPase – the enzyme that acts as the sodium–potassium pump, removing three sodium ions from the cell and taking two potassium ions into the cell during a cycle of action

Transmembrane proteins – proteins that span a membrane and act as channels or transporters of ions

Transporter proteins – a membrane protein involved in the movement of ions, small molecules, and macromolecules (such as another protein) across a membrane

Voltage-gated channels – a class of transmembrane proteins that form ion channels; they are activated by changes in the electrical membrane potential near the channel

Exam-style questions

Structured questions

1 The diagram below shows two conformational states of the sodium–potassium pump in a region of cell membrane.

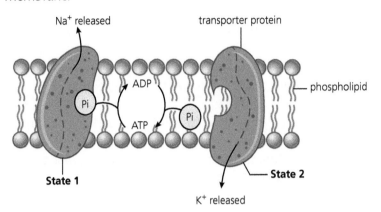

a) Give **one** example of a substance that can pass through the phospholipid layer of a cell membrane. **1**

b) Describe the affinity of the pump for sodium in each of the two states. **2**

c) Give the ratio of sodium and potassium ions pumped in and out of the cell by the action of this pump. **2**

2 The diagram below shows a region of cell membrane containing four protein molecules.

a) Classify the four proteins, R, S, T and V, as integral or peripheral. **2**

b) Describe how protein S is held in place. **1**

c) Name protein R and describe its structure and function in cells. **3**

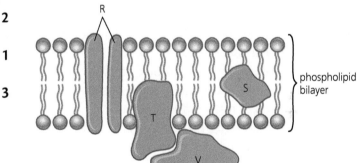

Extended response

3 Give an account of the functioning of gated channel proteins in cell membranes. **5**

4 Give an account of the glucose symport in intestinal cells. **5**

Answers are given on pages 149–150.

Key Area 1.4
Communication and signalling

Key points !

1 Steroid hormones, peptide hormones and neurotransmitters are examples of **extracellular signalling molecules**. ☐

2 Receptor molecules of target cells are proteins with a binding site for a specific signal molecule. ☐

3 Binding of the signal molecules changes the conformation of the receptor, which initiates a response within the cell. ☐

4 Different cell types produce specific signals that can only be detected and responded to by cells with the specific receptor. ☐

5 Signalling molecules may have different effects on different target cell types due to differences in the intracellular signalling molecules and pathways that are involved. ☐

6 In a multicellular organism, different cell types may show a tissue-specific response to the same signal. ☐

7 **Hydrophobic signalling molecules** can diffuse directly through the phospholipid bilayer of membranes, and so bind to intracellular receptors. ☐

8 The receptors for **hydrophobic** signalling molecules are **transcription factors**. ☐

9 Transcription factors are proteins that, when bound to DNA, can either stimulate or inhibit initiation of transcription. ☐

10 The steroid hormones oestrogen and testosterone are examples of hydrophobic signalling molecules. ☐

11 Steroid hormones bind to specific receptors in the cytosol or the nucleus to form a **hormone–receptor complex**. ☐

12 The hormone–receptor complex moves from the cytosol to the nucleus, where it binds to specific sites on DNA and affects gene expression. ☐

13 The hormone–receptor complex binds to specific DNA sequences called **hormone response elements (HREs)** and influences the rate of transcription, with each steroid hormone affecting the gene expression of many different genes. ☐

14 **Hydrophilic signalling molecules** bind to transmembrane receptors and do not enter the cytosol. ☐

15 Peptide hormones and neurotransmitters are examples of **hydrophilic** extracellular signalling molecules. ☐

16 Transmembrane receptors change conformation when the ligand binds to the extracellular face; the signal molecule does not enter the cell, but the signal is transduced across the plasma membrane. ☐

17 Transmembrane receptors allow **signal transduction** by converting the extracellular ligand-binding event into intracellular signals that alter the behaviour of the cell. ☐

18 Transduced hydrophilic signals often involve **G-proteins** or cascades of **phosphorylation** by kinase enzymes. ☐

19 G-proteins relay signals from activated receptors to target proteins such as enzymes and ion channels. ☐

20 **Phosphorylation cascades** allow more than one intracellular signalling pathway to be activated. ☐

21 Phosphorylation cascades involve a series of events with one kinase activating the next in the sequence and so on, resulting in the phosphorylation of many proteins as a result of the original signalling event. ☐

22 Binding of the peptide hormone insulin to its receptor results in an intracellular signalling cascade that triggers recruitment of **GLUT4 glucose transporter proteins** to the cell membrane of fat and muscle cells. ☐

23 Binding of insulin to its receptor causes a conformational change that triggers phosphorylation of the receptor. This starts a phosphorylation cascade inside the cell, which eventually leads to GLUT4-containing vesicles being transported to the cell membrane. ☐

⇨

24 **Diabetes mellitus** can be caused by a failure to produce insulin (type 1) or a loss of receptor function (type 2). ☐

25 Type 2 diabetes is generally associated with obesity. ☐

26 Exercise also triggers recruitment of GLUT4, so can improve uptake of glucose to fat and muscle cells in subjects with type 2 diabetes. ☐

27 The **resting membrane potential** is a state in which there is no net flow of ions across the membrane. ☐

28 The transmission of a nerve impulse requires changes in the membrane potential of the neuron's plasma membrane. ☐

29 An **action potential** is a wave of electrical excitation along a neuron's plasma membrane. ☐

30 Neurotransmitters initiate a response by binding to their receptors at a synapse. ☐

31 Neurotransmitter receptors are ligand-gated ion channels that open when a neurotransmitter binds. ☐

32 **Depolarisation** is a change in the membrane potential to a less negative value inside as positive ions enter. ☐

33 If sufficient ion movement occurs, and the membrane is depolarised beyond a **threshold value**, the opening of voltage-gated sodium channels is triggered and more sodium ions enter the cell down their electrochemical gradient resulting in further depolarisation. ☐

34 A short time after opening, the sodium channels become inactivated, then voltage-gated potassium channels open to allow potassium ions to move out of the cell to restore the resting membrane potential. ☐

35 Depolarisation of a patch of membrane causes neighbouring regions of membrane to depolarise and go through the same cycle, as adjacent voltage-gated sodium channels are opened. ☐

36 When the action potential reaches the end of the neuron it causes vesicles containing neurotransmitter to fuse with the membrane; this releases neurotransmitter, which stimulates a response in a connecting cell. ☐

37 Restoration of the resting membrane potential allows the inactive voltage-gated sodium channels to return to a conformation that allows them to open again in response to depolarisation of the membrane. ☐

38 **Ion concentration gradients** are re-established by the sodium–potassium pump, which actively transports excess ions in and out of the cell. ☐

39 Following **repolarisation** the sodium and potassium ion concentration gradients are reduced. The sodium–potassium pump restores the sodium and potassium ions back to resting potential levels. ☐

40 The retina is the area within the eye that detects light. It contains two types of **photoreceptor cells**: rods and cones. ☐

41 **Rods** function in dim light but do not allow colour perception. **Cones** are responsible for colour vision and only function in bright light. ☐

42 In animals the light-sensitive molecule **retinal** is combined with a membrane protein, **opsin**, to form the photoreceptors of the eye. ☐

43 In rod cells the retinal–opsin complex is called **rhodopsin**. ☐

44 Retinal absorbs a **photon** of light and rhodopsin changes conformation to photoexcited rhodopsin, which starts a cascade of proteins amplifying the signal. ☐

45 A single photoexcited rhodopsin activates hundreds of G-proteins called **transducins**, which activate a single molecule of the enzyme **phosphodiesterase (PDE)** each. ☐

46 PDE catalyses the hydrolysis of a molecule called **cyclic GMP (cGMP)**. ☐

47 Each active PDE molecule breaks down thousands of cGMP molecules per second. ☐

48 The reduction in cGMP concentration as a result of its hydrolysis affects the function of ion channels in the membrane of rod cells, resulting in the closure of ion channels and triggering nerve impulses in neurons in the retina. ☐

49 A very high degree of amplification results in rod cells being able to respond to low intensities of light. ☐

50 In cone cells, different forms of opsin combine with retinal to give different photoreceptor proteins, each with a maximal sensitivity to specific wavelengths: red, green, blue or UV. ☐

Co-ordination

Multicellular organisms have many cells and they signal to each other using extracellular signalling molecules to ensure the co-ordinated functioning of the whole organism. Steroid hormones, peptide hormones and neurotransmitters are examples of extracellular signalling molecules.

The signal molecules released from one cell are specific to the receptor molecules of the target cells. Receptor molecules are proteins with a binding site for a specific signal molecule. Binding changes the conformation of the receptor, which initiates a response within the target cell.

Different cell types produce specific signals that can only be detected and responded to by other cells with the specific receptor. Signalling molecules may have different effects on different target cell types due to differences in the intracellular signalling molecules and pathways that are involved. In a multicellular organism, different cell types may show a tissue-specific response to the same signal.

Hydrophobic signal molecules

Hydrophobic signalling molecules can diffuse directly through the phospholipid bilayers of membranes, and so bind to intracellular receptors located either in the cytosol or nucleus.

The receptors for hydrophobic signalling molecules are transcription factors. These transcription factors are proteins that, when bound to DNA, can either stimulate or inhibit the initiation of transcription. The steroid hormones oestrogen and testosterone are examples of hydrophobic signalling molecules. Steroid hormones bind to specific receptors in the cytosol or the nucleus.

Cytosolic hormone–receptor complexes move to the nucleus where they bind to specific sites on DNA and affect gene expression. The specific DNA sequences at these sites are called hormone response elements (HREs). Binding at these sites influences the rate of transcription, with each steroid hormone generally affecting the gene expression of many different genes, as shown in Figure 1.38.

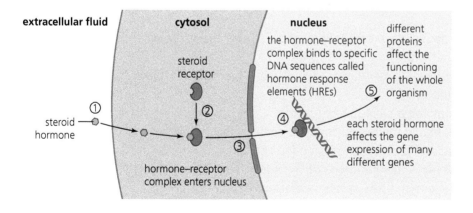

Figure 1.38 ① Steroid hormone enters cell through phospholipid membrane; ② it binds to specific intracellular receptor; ③ hormone–receptor complex enters nucleus; ④ it binds to HRE where transcription is affected; ⑤ the different protein produced, or not produced, affects the functioning of the whole organism

---

Check-up 17 ❓

1   Describe how hydrophobic signal molecules can enter cells.   **1**
2   Describe how hydrophobic signal molecules can affect gene expression.   **3**

Hydrophilic signal molecules

Hydrophilic signalling molecules bind to transmembrane receptors as ligands and do not enter the cytosol. Peptide hormones and neurotransmitters are examples of hydrophilic extracellular signalling molecules. Transmembrane receptors change conformation when the ligand binds to the extracellular face and act as signal transducers. Although the signal molecule does not enter the cell, the signal is transduced across the plasma membrane and the extracellular ligand-binding event is converted into intracellular signals, which alter the behaviour of the cell.

Signal transduction

Transduced hydrophilic signals often involve G-proteins or cascades of phosphorylation by kinase enzymes. G-proteins relay signals from activated receptors that have bound to a signalling molecule to target proteins such as enzymes and ion channels. Phosphorylation cascades allow more than one intracellular signalling pathway to be activated. Phosphorylation cascades involve a series of events with one kinase activating the next in the sequence, and so on. Phosphorylation cascades can result in the phosphorylation of many proteins as a result of the original signalling event.

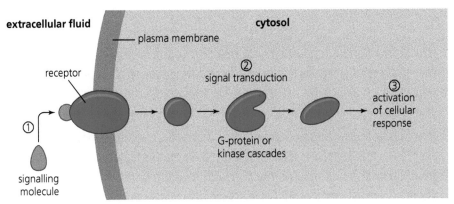

Figure 1.39 Transduction of a hydrophilic signal ① signal molecules binds to receptor; ② signal transduced through a G-protein or kinase cascade; ③ cellular response activated

Check-up 18 ?

1   Describe how hydrophilic signal molecules act specifically.   2
2   Describe the process of signal transduction.   2

Example

Control of blood glucose by the action of insulin

Elevation of blood glucose levels in humans cause the release of the peptide hormone insulin from cells in the pancreas. The binding of insulin to its receptor in the membranes of fat and muscle cells causes a conformational change that triggers phosphorylation of the receptor. This starts a phosphorylation cascade inside the cell, which eventually leads to GLUT4-containing vesicles being translocated and recruited into the cell membrane, as shown in Figure 1.40. The GLUT4s allow entry of glucose into the cell where it can be converted to glycogen or used in glycolysis, thus lowering blood glucose concentration.

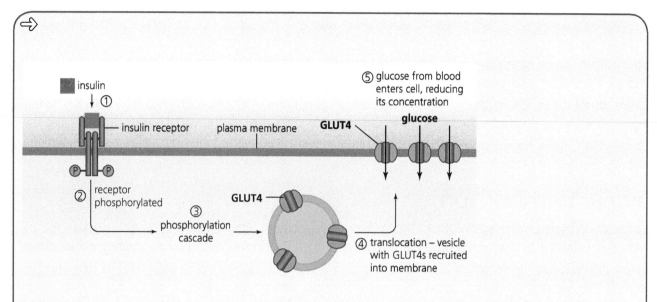

Figure 1.40 ① Insulin molecule binds to receptor; ② receptor phosphorylated; ③ cascade of phosphorylation; ④ vesicle with GLUT4s becomes recruited into membrane; ⑤ glucose taken up by cell via GLUT4s, lowering glucose concentration in blood

Diabetes mellitus

Diabetes mellitus is an illness in which the signalling system that helps to regulate blood glucose concentrations fails. Type 1 diabetes is present from birth and is caused by a failure to produce the peptide hormone insulin. The events shown in Figure 1.40 cannot be triggered, so blood glucose remains high following a meal. Type 1 diabetes is treated with targeted doses of injected insulin.

Type 2 diabetes begins later in life and is generally associated with obesity. It is caused by the loss of insulin receptor function, which fails to trigger the events in Figure 1.40 even in the presence of insulin. Type 2 diabetes is treated with some drugs but mainly through lifestyle changes related to diet and exercise levels. Exercise triggers recruitment of GLUT4, so can improve uptake of glucose to fat and muscle cells in people with type 2 diabetes.

Check-up 19

1 Explain how insulin can cause a decrease in blood glucose concentration.   **4**
2 Describe the differences between type 1 and type 2 diabetes.   **4**

Nerve impulse transmission

Initiating an action potential

The resting membrane potential of a neuron is a state in which there is no nervous impulse being transmitted. The inside of the membrane is more negative compared to the outside, which is relatively more positive. The transmission of a nerve impulse requires changes in the membrane potential of the neuron's plasma membrane.

Neurotransmitter receptors are ligand-gated ion channels in the synaptic region at one end of a neuron. Neurotransmitter substances act as ligands that bind to these gates. Binding causes the channels to open and sodium ions enter the cell as shown in Figure 1.41a). If sufficient ion movement through these channels occurs, and the membrane is depolarised by becoming less negative inside beyond a threshold value, the opening of voltage-gated sodium channels is triggered and sodium ions enter the cell down their electrochemical gradient, causing further depolarisation to occur, shown in Figure 1.41b).

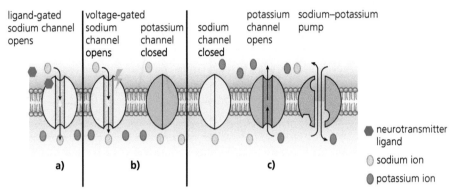

a)   b)   c)

neurotransmitter ligand

sodium ion

potassium ion

Figure 1.41 a) Binding of neurotransmitter to a ligand-gated channel opens the channel and allows sodium ions to enter; b) changes in potential caused by the entry of sodium ions opens voltage-gated sodium channels, which allows entry of more sodium ions, maybe reaching threshold level; c) changes in potential now close sodium channels and open potassium channels, causing repolarisation; the sodium–potassium pump restores the resting potential

This rapidly leads to a large change in the membrane potential, causing an action potential to be generated in this region of the neuron. Depolarisation of this first region of neuron causes the next region to depolarise and go through the same cycle of events, allowing the action potential to travel rapidly along the neuron, region after region. This is the nerve impulse as shown in Figure 1.42.

Restoration of resting potential

It is crucial that a neuron goes back to its resting potential rapidly. Restoration of the resting membrane potential allows the inactive voltage-gated sodium channels to return to a conformation that allows them to open again in response to further signals, allowing the system to remain sensitive.

A short time after opening, the sodium channels become inactivated, stopping movement of sodium ions into the cell. Voltage-gated potassium channels then open to allow potassium ions to move out of the cell to restore the resting membrane potential. Ion concentration gradients are re-established by the sodium–potassium pump, which actively transports excess ions in and out of the cell as shown in Figure 1.41c).

Check-up 20 ❓

1 Describe how an action potential is achieved in neurons.   4
2 Describe how the resting potential is restored following the passage of an action potential.   4

Transmission at a synapse

When the action potential reaches the end of the neuron it causes vesicles containing neurotransmitter to fuse with the membrane, releasing neurotransmitter into the next synaptic cleft, which stimulates a response in the next connected cell.

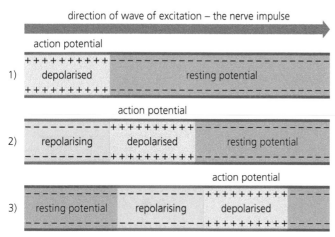

direction of wave of excitation – the nerve impulse

1)
action potential
+ + + + + + + + + – – – – – – – – – – – – – – – – –
depolarised   resting potential
+ + + + + + + + + – – – – – – – – – – – – – – – – –

2)
action potential
– – – – – – – – – + + + + + + + + + – – – – – – – –
repolarising   depolarised   resting potential
– – – – – – – – – + + + + + + + + + – – – – – – – –

3)
action potential
– – – – – – – – – – – – – – – – – – + + + + + + + + + – – – – –
resting potential   repolarising   depolarised
– – – – – – – – – – – – – – – – – – + + + + + + + + + – – – – –

Figure 1.42 Transmission of a nerve impulse along a neuron: 1) ligand-gated and voltage-gated ion channels depolarise the membrane by admitting sodium ions into the neuron, reaching a threshold level for transmission; 2) sodium channels close and potassium channels open, causing repolarisation of the membrane; 3) the sodium–potassium pump helps restore resting potential; this series of changes is repeated region after region along the neuron, causing the transmission of the impulse followed by the restoration of the resting potential

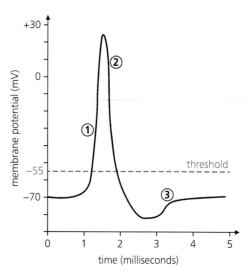

Figure 1.43 Electrical changes in a region of neuron during the transmission of a nerve impulse: ① depolarisation as sodium ions enter; ② repolarisation as potassium channels open; ③ resting potential restored as sodium–potassium pump is active

Example

The vertebrate eye

The retina is the area within the vertebrate eye that detects light. It contains two types of photoreceptor cells – rods and cones – as shown in Figure 1.44. Rods function in dim light but do not allow colour perception. Cones are responsible for colour vision and only function in bright light.

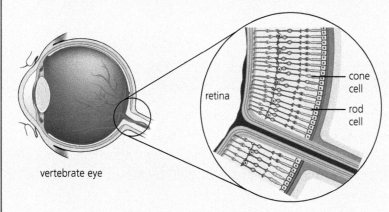

Figure 1.44 Vertebrate eye showing details of the cells in the retina

Photoreceptors

In animals, the light-sensitive molecule retinal is combined with a membrane protein, opsin, to form the photoreceptors of the eye, as shown in Figure 1.45.

In rod cells the retinal–opsin complex is called rhodopsin. Retinal absorbs a photon of light and rhodopsin changes conformation to form photoexcited rhodopsin. Photoexcited rhodopsin activates hundreds of G-protein molecules called transducins, each of which activates one molecule of the enzyme phosphodiesterase

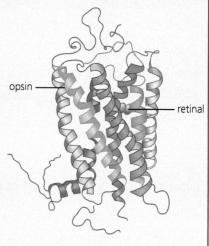

Figure 1.45 Rhodopsin molecule

(PDE). PDE catalyses the hydrolysis (breakdown) of a molecule called cyclic GMP (cGMP). Each active PDE molecule breaks down thousands of cGMP molecules per second. The reduction in cGMP concentration affects the function of ion channels in the membrane of rod cells, which close to trigger nerve impulses in neurons in the retina, as shown in Figure 1.46. This means that a very high degree of amplification has resulted and rod cells can respond to low intensities of light.

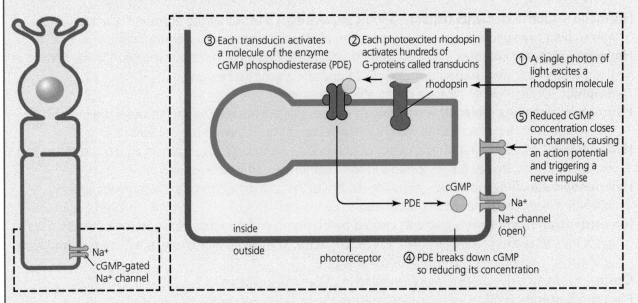

Figure 1.46 Rod cell with magnified region to show amplification of the stimulus from one photon to trigger a nervous impulse in a neuron

In cone cells, different forms of opsin combine with retinal to give different photoreceptor proteins, each with a maximal sensitivity to specific wavelengths: red, green, blue or UV.

Check-up 21 ?

1 Describe the structure of rhodopsin. **2**
2 Describe how a photon of light has its effect amplified in a rod cell to produce a nervous impulse. **4**

Key words

Action potential – a wave of electrical excitation along a neuron's plasma membrane

Cones – photoreceptor cells responsible for colour vision; they only function in bright light

Cyclic GMP (cGMP) – a second messenger for visual transduction; it is present in high concentrations in photoreceptor cells

Depolarisation – an electrical state in an excitable cell whereby the inside of the cell is made less negative relative to the outside than it is at the resting membrane potential

Diabetes mellitus – an inability to regulate blood glucose levels; in type 1 there is a failure to produce insulin; in type 2 there is a loss of function of insulin receptors on cell surface

Extracellular signalling molecule – cues (such as growth factors, hormones, cytokines and neurotransmitters) designed to transmit specific information to target cells

GLUT4 glucose transporter proteins – the insulin-regulated glucose transporter; insulin triggers the movement of GLUT4 transporters to the membrane surface, increasing uptake of glucose to be converted to glycogen

G-proteins – also known as guanine nucleotide-binding proteins; a family of proteins that act as molecular switches inside cells, which are involved in transmitting signals from a variety of stimuli outside a cell to its interior

Hormone response elements (HREs) – a short sequence of DNA within the promoter of a gene that is able to bind to a specific hormone–receptor complex and therefore regulate transcription

Hormone–receptor complex – formed when steroid hormones bind to specific receptors in the cytosol or the nucleus; they bind to specific sites on DNA and affect gene expression

Hydrophilic – 'water loving'; having a strong affinity for water

Hydrophilic signalling molecule – signalling molecules that are not able to pass through the membrane; the signal is transduced across the membrane by receptor molecules on the cell surface

Hydrophobic – 'water-fearing'; the tendency of non-polar substances to aggregate in aqueous solution and exclude water molecules; it is responsible for cell membrane and vesicle formation

Hydrophobic signalling molecule – signalling molecules that can diffuse through membranes, so their receptor molecules can be within the nucleus

Ion concentration gradient – gradients created by ion pump enzymes that transport ionic solutes, such as sodium, potassium, hydrogen ions and calcium, across the membrane; energy is required to produce a gradient, so the gradient is a form of stored energy

Opsin – a light-sensitive protein molecule found in the animal kingdom

Phosphodiesterase (PDE) – enzyme that catalyses the hydrolysis of cyclic GMP (cGMP)

Phosphorylation – the addition of a phosphate group to a protein or other organic molecule

Phosphorylation cascades – a series of events in which one kinase activates the next one in a sequence; phosphorylation cascades can result in the phosphorylation of many proteins as a result of the original signalling event

Photon – the basic unit of light

Photoreceptor cells – cells (rods and cones) found in the retina that are capable of visual phototransduction (converting light – visible electromagnetic radiation – into electrical signals)

Repolarisation – the restoration of a membrane potential following depolarisation (that is, restoring a negative internal charge)

Resting membrane potential – the difference in ion concentration between the inside and outside of a cell

Retinal – a light-sensitive molecule within the eye that binds with opsin to form photoreceptors in the eye

Rhodopsin – the retinal–opsin complex in rod cells

Rods – photoreceptor cells in the retina that function in dim light and respond to low light intensities; they do not allow colour perception

Signal transduction – conversion of extracellular signals into an intracellular response

Threshold value – when the depolarisation reaches about -55 mV, a neuron will fire an action potential – this is the threshold; if the neuron does not reach this critical threshold level, the action potential will not fire

Transcription factor – a protein that binds to specific DNA sequences, thereby controlling the rate of transcription of genetic information from DNA to messenger RNA

Transducin – a protein naturally expressed in vertebrate retina rods and cones; it is very important in vertebrate phototransduction

Exam-style questions

Structured questions

1   The diagram opposite shows a rod cell from a vertebrate eye.
   a) Name the components of a rhodopsin molecule
      from a vertebrate eye.   2
   b) When a photon of light strikes a rhodopsin
      molecule, transducin proteins are activated.
      Describe the effects of activated
      transducins.   2
   c) Describe the effects of reduced levels of
      cyclic GMP on sodium ion channels in the
      rod cell.   1
   d) Rods cells amplify the effects of light.
      State why amplification is an advantage to
      vertebrates.   1
   e) State how the photoreceptor molecules
      of cone cells differ from those in
      rod cells.   1
   f) Explain why having a variety of different cone
      cells is an advantage to vertebrates.   1

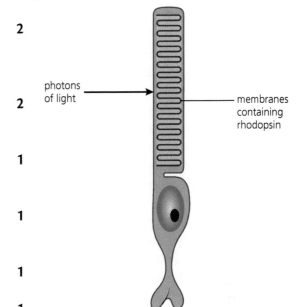

2   The diagram below shows changes in electrical activity in a region of a neuron as a nervous impulse
    passes.

   a) Describe how the binding of a neurotransmitter
      molecule can start to change the resting
      potential of the membrane as shown at X
      on the graph.   2
   b) Describe how voltage-gated channels are
      activated and their effects on the resting
      potential.   2
   c) State what is meant by the 'threshold' shown
      on the graph.   1
   d) Explain how the repolarisation of the membrane
      is achieved after the impulse has passed at Y on
      the graph.   2
   e) Explain why it is important that the neuron
      is restored to resting potential following the
      transmission of a nervous impulse.   1

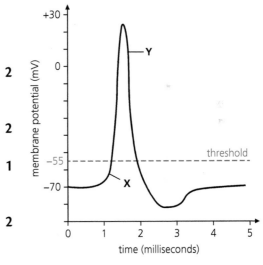

Extended response

3   Give an account of the role of insulin in the decrease in glucose concentration in blood.   4
4   Give an account of hydrophobic cell signalling.   6

Answers are given on pages 150–151.

Key Area 1.5
Protein control of cell division

Key points !

1   The **cytoskeleton** gives mechanical support and shape to cells. ☐
2   The cytoskeleton consists of different protein structures including **microtubules**, which are found in all eukaryotic cells. ☐
3   Microtubules are hollow cylinders, composed of the protein **tubulin**, which radiate from the **microtubule organising centre (MTOC)**, or centrosome. ☐
4   Microtubules control the movement of membrane-bound organelles and chromosomes. ☐
5   Cell division requires remodelling of the cytoskeleton. ☐
6   Formation and breakdown of microtubules involves the polymerisation and depolymerisation of tubulin. ☐
7   Microtubules form the **spindle fibres** that are active during cell division. ☐
8   The **cell cycle** consists of **interphase** and the mitotic (M) phase. ☐
9   Interphase involves growth and DNA synthesis including:
    a) G1, a growth phase
    b) S, a phase during which the DNA is replicated
    c) G2, a further growth phase. ☐
10 The mitotic phase involves **mitosis** and **cytokinesis**. In mitosis the chromosomal material is separated by the spindle microtubules. This is followed by cytokinesis, in which the cytoplasm is separated into two daughter cells. ☐
11 Mitosis consists of **prophase**, **metaphase**, **anaphase** and **telophase**. ☐
12 In prophase, DNA condenses into chromosomes, each consisting of two sister chromatids; the nuclear membrane breaks down; and spindle microtubules extend from the MTOC by polymerisation and attach to chromosomes via their **kinetochores** in the **centromere** region. ☐
13 In metaphase, chromosomes are aligned at the metaphase plate (equator of the spindle). ☐
14 In anaphase, spindle microtubules shorten by depolymerisation, sister chromatids are separated, and the chromosomes are pulled to opposite poles. ☐
15 In telophase, the chromosomes decondense and nuclear membranes are formed around them. ☐
16 Progression through the cell cycle is controlled by **cell cycle checkpoints**, which are mechanisms within the cell that assess the condition of the cell during the cell cycle and halt progression to the next phase until certain requirements are met. ☐
17 **Cyclin proteins** that accumulate during cell growth are involved in regulating the cell cycle. ☐
18 Cyclins combine with and activate **cyclin-dependent kinases (CDKs)**. ☐
19 Active cyclin–CDK complexes phosphorylate proteins that regulate progression through the cycle; if sufficient phosphorylation is reached, progression occurs. ☐
20 At the G1 checkpoint, **retinoblastoma protein (Rb)** acts as a tumour suppressor by inhibiting the transcription of genes that code for proteins needed for DNA replication. ☐
21 Phosphorylation by **G1 cyclin–CDK** inhibits Rb. ☐
22 Inhibition of Rb allows transcription of the genes that code for proteins needed for DNA replication, and cells progress from G1 to S phase. ☐
23 At the G2 checkpoint, the success of DNA replication and any damage to DNA is assessed. ☐
24 DNA damage triggers the activation of several proteins, including **p53**, that can stimulate DNA repair, arrest the cell cycle or cause cell death. ☐
25 A metaphase checkpoint controls progression from metaphase to anaphase. ☐
26 At the metaphase checkpoint, progression is halted until the chromosomes are aligned correctly on the metaphase plate and attached to the spindle microtubules. ☐

⇨

⇨

27 An uncontrolled reduction in the rate of the cell cycle may result in degenerative disease. ☐

28 An uncontrolled increase in the rate of the cell cycle may result in tumour formation. ☐

29 A **proto-oncogene** is a normal gene, usually involved in the control of cell growth or division, that can mutate to form a **tumour-promoting oncogene**. ☐

30 **Apoptosis** is programmed cell death which is triggered by cell death signals that can be external or internal. ☐

31 The production of **death signal molecules** from lymphocytes is an example of an external death signal. ☐

32 External death signal molecules bind to a surface receptor protein and trigger a protein cascade within the cytoplasm. ☐

33 An internal death signal resulting from DNA damage causes activation of the p53 tumour-suppressor protein. ☐

34 Both types of death signal result in the activation of **caspase cascades**, which cause the destruction of the cell. ☐

35 Apoptosis is essential during the development of an organism to remove cells that are no longer required as development progresses or during **metamorphosis**. ☐

36 Cells may initiate apoptosis in the absence of **growth factors**. ☐

Cell division and the remodelling of the cytoskeleton

The cytoskeleton is a fibrous framework that gives mechanical support and shape to cells. The cytoskeleton consists of different protein structures including microtubules, which are found in all eukaryotic cells, as shown in Figure 1.47a). Microtubules are hollow cylinders composed of molecules of the protein tubulin, as shown in Figure 1.47b). They radiate from the microtubule organising centre (MTOC), or centrosome. Microtubules control the movement of membrane-bound organelles that are attached to them and the movements of chromosomes during cell division.

Cell division requires remodelling of the cytoskeleton. The formation of microtubules during remodelling involves polymerisation of tubulin by adding more units. Breakdown of microtubules occurs due to the depolymerisation of tubulin, as shown in Figure 1.48a). Microtubules form the spindle fibres that are active during cell division, as shown in Figure 1.48b).

a)
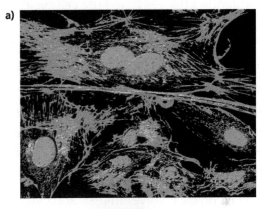

b)
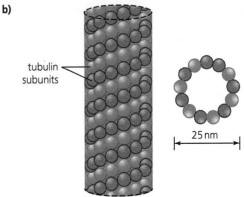

Figure 1.47 a) The green fibres in this fluorescently labelled image represent microtubules in these eukaryotic cells; b) the structure of a microtubule showing the subunits of tubulin that make up their hollow structure and their fine diameter

Key links 👍

There is more about fluorescence microscopy in Key Area 1.1.

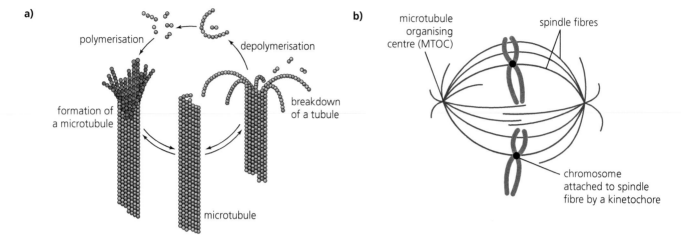

Figure 1.48 The restructuring of the cytoskeleton involves the polymerisation and depolymerisation of tubulin in microtubules; b) spindle in a dividing cell showing chromosomes attached to the fibres by kinetochores

The cell cycle

During the active growth and development of eukaryotic organisms, individual cells undergo sequential changes called the cell cycle. In the mature organism, many cells no longer growing or dividing enter a final resting stage in the body of the organism. The cell cycle consists of interphase and the mitotic (M) phase, as shown in Figure 1.49.

Interphase involves growth and DNA synthesis, including G1, a growth phase; the S phase, during which the DNA is replicated; and G2, a further growth phase in preparation for mitosis.

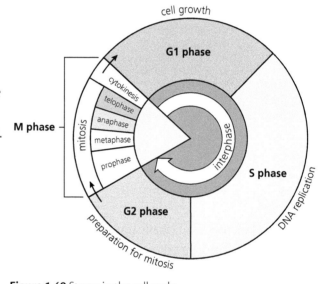

Figure 1.49 Stages in the cell cycle

The M phase involves mitosis and cytokinesis. In mitosis the chromosomal material is separated by the spindle microtubules. This separation is followed by cytokinesis, in which the cytoplasm is separated into two daughter cells.

Mitosis

Mitosis consists of prophase, metaphase, anaphase and telophase, as shown in Figure 1.50.

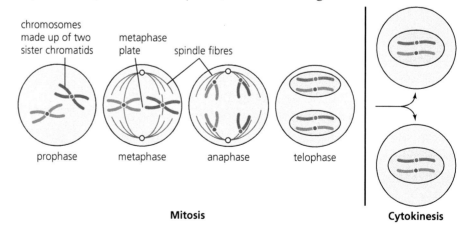

Figure 1.50 Mitotic stages of the cell cycle and cytokinesis

Mitosis **Cytokinesis**

- **Prophase** – DNA condenses into chromosomes, each consisting of two sister chromatids. The nuclear membrane breaks down, and spindle microtubules extend from the MTOC by polymerisation and attach to the chromosomes via their kinetochores in the centromere region.

- **Metaphase** – chromosomes are aligned at the metaphase plate or equator of the spindle.
- **Anaphase** – as spindle microtubules shorten by depolymerisation, sister chromatids are separated, and the chromosomes formed are pulled to opposite poles.
- **Telophase** – the chromosomes start to decondense and new nuclear membranes are formed around them.

Hints & tips

Make sure you know the difference between a chromosome and a chromatid – it's a bit confusing! Chromatids are replicated chromosomes held together at a centromere; when they are pulled apart during anaphase, each becomes a chromosome in its own right.

Check-up 22 ?

1 Describe the role of the cytoskeleton and describe how it can be remodelled during cell division.	**2**
2 Name the different phases of a normal cell cycle and state what occurs in each.	**4**
3 Describe the movements of the chromosomes during mitosis.	**4**

Control of the cell cycle

Progression through the cell cycle is regulated by checkpoints at G1, G2 and metaphase in mitosis, as shown in Figure 1.51. Checkpoints are mechanisms within the cell that assess its condition during the cell cycle and halt progression to the next phase until certain requirements are met. If a go-ahead signal is not reached at the G1 checkpoint, the cell may switch to a non-dividing or resting state called G0.

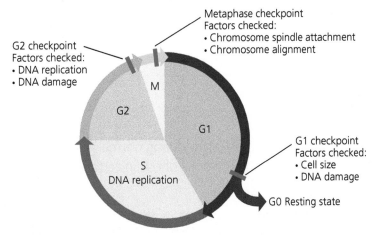

Figure 1.51 Checkpoint control of the cell cycle with the main factors being checked

Cyclin proteins that accumulate during cell growth are involved in regulating the cell cycle. Cyclins combine with and activate cyclin-dependent kinases (CDKs). Active cyclin–CDK complexes phosphorylate proteins that regulate progression through the cycle. If sufficient phosphorylation is reached, progression occurs.

G1 checkpoint

At the G1 checkpoint, the size of the cell is checked to confirm that there is sufficient cell mass for daughter cells to be produced. Retinoblastoma protein (Rb) acts as a tumour suppressor in G1 by inhibiting the transcription of genes that code for proteins needed for DNA replication. Phosphorylation by G1 cyclin–CDK inhibits Rb, which allows transcription of the genes that code for proteins such as enzymes needed for DNA replication, as shown in Figure 1.52. If this happens, cells progress from G1 to the S phase where DNA replication occurs.

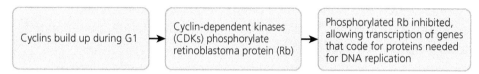

Figure 1.52 Events that allow go-ahead at G1 and take the cell into the S phase of the cell cycle

DNA damage is also monitored at G1. Damage triggers the activation of several proteins, including p53, which can stimulate DNA repair, arrest the cell cycle or cause cell death.

G2 checkpoint

At the G2 checkpoint, the success of DNA replication and any damage to DNA is assessed. From here cells progress to the M phase, where mitosis occurs.

Metaphase checkpoint

A metaphase checkpoint controls progression from metaphase to anaphase. At the metaphase checkpoint, progression is halted until the chromosomes are aligned correctly on the metaphase plate and securely attached to the spindle microtubules. This checkpoint controls the entry to anaphase.

Tumour formation

An uncontrolled increase in the rate of the cell cycle may result in tumour formation. A proto-oncogene is a normal gene, usually involved in the control of cell growth or division, that can mutate to form a tumour-promoting oncogene.

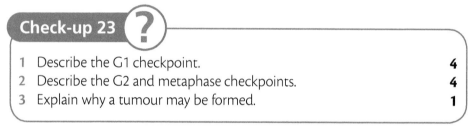

Check-up 23 ❓

1	Describe the G1 checkpoint.	**4**
2	Describe the G2 and metaphase checkpoints.	**4**
3	Explain why a tumour may be formed.	**1**

Apoptosis

Apoptosis is programmed cell death, which can occur during normal growth and development, resulting in the removal of old or damaged cells or during metamorphosis in certain species, as shown in Figure 1.53.

Apoptosis can also be used to kill cells that have started to divide in an uncontrolled way during tumour formation. Apoptosis is triggered by cell death signals, which can be external or internal:

- The production of death signal molecules from lymphocytes is an example of an external death signal. External death signal molecules bind to a surface receptor protein and trigger a protein cascade within the cytoplasm.
- DNA damage is an example of an internal death signal. An internal death signal resulting from DNA damage causes activation of p53 tumour-suppressor protein.

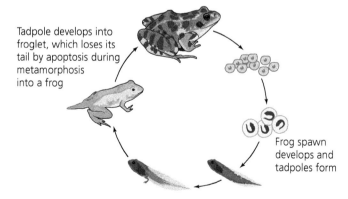

Figure 1.53 Metamorphosis in frogs – loss of cells in the tail of froglets is due to apoptosis

Both types of death signal result in the activation of protease enzymes called caspases, which act in cascades to cause the destruction of the cell, as shown in Figure 1.54.

Cells may initiate apoptosis in the absence of growth factors.

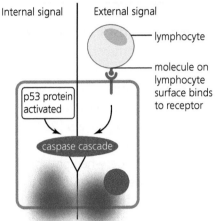

Figure 1.54 Internal or external death signals can trigger apoptosis

Check-up 24 ?

1 Explain how apoptosis can be triggered.   **3**
2 Explain the significance of apoptosis for the whole organism. **2**
3 Explain why a tumour may be formed.   **1**

Key words

Anaphase – phase in which spindle microtubules shorten by depolymerisation, sister chromatids are separated, and the chromosomes are pulled to opposite poles

Apoptosis – programmed cell death triggered by cell death signals that can be external or internal

Caspase cascade – protease enzymes involved in a series of reactions (a cascade) that destroy a cell

Cell cycle – four-stage process in which the cell increases in size (G1 stage), copies its DNA (S stage), prepares to divide (G2 stage) and divides by mitosis (M stage)

Cell cycle checkpoints – checkpoints during G1, G2 and metaphase that assess the readiness of a cell to enter the next stage of the cell cycle

Centromere – the specialised DNA sequence of a chromosome that links a pair of sister chromatids

Cyclin proteins – proteins that control the progression of cells through the cell cycle by activating cyclin-dependent kinase (CDK) enzymes

Cyclin-dependent kinases (CDKs) – when activated by cyclin, CDKs cause the phosphorylation of proteins, which stimulates the cell cycle

Cytokinesis – division of cytoplasm to form two daughter cells

Cytoskeleton – a microscopic network of protein filaments and tubules in the cytoplasm of many living cells that supports their shape and function

Death signal molecules – external and internal signals that result in the activation of protease enzymes called caspases, which cause apoptosis

G1 cyclin–CDK – phosphorylation by G1 cyclin–CDK inhibits the retinoblastoma protein (Rb); this allows the transcription of the genes that code for proteins needed for DNA replication, allowing cells to progress from G1 to S phase

Growth factors – a naturally occurring substance, usually a protein or steroid hormone, capable of stimulating cellular growth, proliferation healing, and cellular differentiation

Interphase – phase of the cell cycle in which the cell spends the majority of its time; it consists of the G1, S and G2 phases in preparation for the M phase

Kinetochores – a complex of proteins associated with the centromere of a chromosome during cell division, to which the microtubules of the spindle attach

Metamorphosis – process that involves a significant change in an organism's physical form during development

Metaphase – phase in which chromosomes align at the metaphase plate (equator of the spindle) and attach to the spindle fibres

Microtubule organising centre (MTOC) – structure found in eukaryotic cells from which microtubules are produced for the formation of the spindle fibres

Microtubules – microscopic hollow tubes made of the protein tubulin that are part of a cell's cytoskeleton

Mitosis – division of the nucleus to form two new nuclei, each with a full complement of chromosomes

p53 – a tumour-suppressor protein that can stimulate DNA repair, arrest the cell cycle or cause cell death (apoptosis) by the activation of caspases   ⇒

Prophase – phase in which DNA condenses into chromosomes, each consisting of two sister chromatids; the nuclear membrane breaks down, the spindle forms and chromosomes attach via their kinetochores in their centromere region

Proto-oncogene – a normal gene, usually involved in the control of cell growth or division, which can mutate to form a tumour-promoting oncogene

Retinoblastoma protein (Rb) – a tumour-suppressor protein that is dysfunctional in several major cancers; one function of Rb is to prevent excessive cell growth by inhibiting cell cycle progression until a cell is ready to divide; when phosphorylated, it allows DNA replication in the S phase

Spindle fibres – microtubules to which chromosomes are attached by kinetochores during cell division

Telophase – phase in which the chromosomes decondense and nuclear membranes are formed around them

Tubulin – the protein that polymerises into long chains or filaments that form microtubules, which serve as a cytoskeleton for living cells

Tumour-promoting oncogene – a mutated proto-oncogene gene that has the potential to cause cancer

Exam-style questions

Structured questions

1 The **diagram** below represents the cell cycle, which consists of interphase (G1, S and G2), mitosis and cytokinesis (M). The **flowchart** shows events that lead up to the G1 checkpoint.

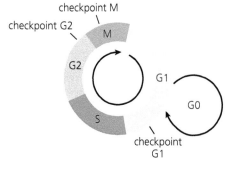

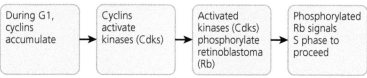

During G1, cyclins accumulate → Cyclins activate kinases (Cdks) → Activated kinases (Cdks) phosphorylate retinoblastoma (Rb) → Phosphorylated Rb signals S phase to proceed

a) (i) Give the probable results of uncontrolled increases and decreases in the rate of the cell cycle. **2**

(ii) Explain why cells may enter G0, as shown in the diagram. **2**

(iii) Describe what happened in the S phase. **1**

(iv) Describe how the cytoskeleton is remodelled during cell division. **2**

(v) State what happens to a cell during cytokinesis at the end of the M phase shown on the diagram. **1**

b) (i) Give **one** change in a cell as it moves through G1 leading to the accumulation of cyclins shown in the flowchart. **1**

(ii) Explain how retinoblastoma (Rb) in the flowchart can act as a tumour suppressor in G1. **3**

Extended response

2 Give an account of the different cell death signals and their effects on cells. **5**

3 Give an account of the events of the different phases of mitosis. **10**

Answers are given on page 151.

Practice course assessment: cells and proteins

Section 1

1   Which of the following techniques would be used to separate target proteins from a mixture of proteins?

   A immunoassay   C electrophoresis

   B colorimetry   D thin layer chromatography

2   Propidium iodide is used as a vital stain to directly identify viable cells when viewed in a haemocytometer. Vital stains such as propidium iodide

   A stain all cells   C only stain living cells

   B stain all dead cells   D only stain the culture medium

3   The diagrams below represent the general actions of enzymes involved in the transfer of phosphate groups in cells.

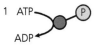 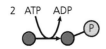 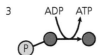

Key

● Protein

Ⓟ Phosphate group

	Phosphatases	ATPases	Kinases
A	1	2	3
B	1	3	2
C	2	3	1
D	3	1	2

Which line in the table identifies the enzymes in each case?

4   Which line in the table represents the binding site and effect on affinity of an allosteric enzyme binding with a positive modulator?

	Modulator binding site		Affinity of enzyme for substrate	
	Active site	Allosteric site	Increased	Decreased
A	✓		✓	
B		✓		✓
C		✓	✓	
D	✓			✓

5   The diagram below shows the arrangement of four proteins (R, S, T and V) and the phospholipid bilayer of a cell membrane.

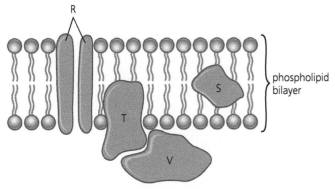

Which of the proteins shown interact extensively with the hydrophobic region of membrane phospholipids?

   A S only   B V only   C R, S and T only   D R, S, T and V

6 The diagram below shows the action of a glucose and sodium ion (Na⁺) symport into a cell from the lining of the human small intestine. The transport of glucose into this cell is facilitated by the symport and the transport of Na^+ is down its concentration gradient.

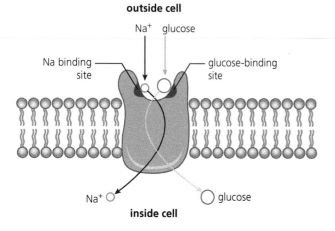

Which line in the table correctly represents the relative concentrations of glucose and sodium ions (Na^+) on the two sides of the plasma membrane at the start of the process?

	Sodium ion (Na^+)		Glucose	
	Outside cell	Inside cell	Outside cell	Inside cell
A	High	Low	Low	High
B	High	Low	High	Low
C	Low	High	Low	High
D	Low	High	High	Low

7 Oestrogen and testosterone are steroid hormones.
Which letter in the diagram below shows correctly the movement by molecules of these hormones in the first stage in its cell signalling process?

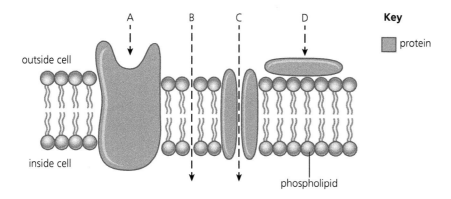

8 Which line in the table correctly identifies features of cone cells from human eyes?

	Able to function in low light intensity	Contain different forms of opsin in their photoreceptor proteins compared to rod cells	Photoreceptor protein contains retinal
A	Yes	Yes	No
B	Yes	No	Yes
C	No	No	No
D	No	Yes	Yes

9 Which of the following diagrams correctly represents the sequence of phases in the cell cycle?

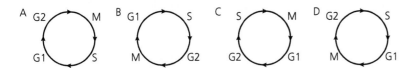

10 Retinoblastoma protein (Rb) has a role in the regulation of progress through the cell cycle. It can be phosphorylated (Rb-P) or unphosphorylated (Rb).
Which line in the table shows the phase in the cell cycle during which Rb functions as a regulator, and which of its phosphorylation states allows the cell cycle to progress?

	Phase	Phosphorylation state which allows the cycle to progress
A	G1	phosphorylated
B	G1	not phosphorylated
C	S	phosphorylated
D	S	not phosphorylated

Section 2

1 The diagram below shows the final stage in a test that confirms a blood sample contains antibodies against the herpes simplex virus (HSV). HSV antigen is attached to the plastic well and any unbound areas are coated with non-reactive material.

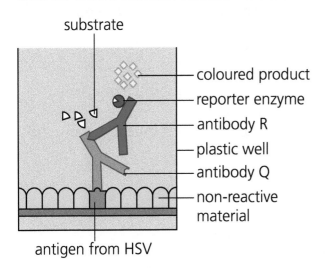

a) Identify the technique represented in the diagram.   1
b) Antibody Q is a monoclonal antibody. Give the meaning of the term 'monoclonal'.   1
c) Explain why the test represented would not reveal if the person had been infected with
   chickenpox virus.   1
d) Use information from the diagram to explain why inadequate rinsing just before the
   stage shown might result in a false positive result.   1
e) A pH buffer was used in all reagents and wash solutions.
   Explain why it is important to control pH.   1
f) Antibody R is linked to a reporter enzyme. Give **one** other example of a chemical
   'label' that can be used in this technique.   1

2 The diagrams below represent the molecular structure of the proteins haemoglobin and histone 4 (H4).

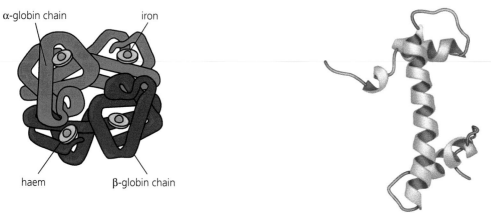

Haemoglobin   Histone 4 (H4)

a) Describe what is meant by the primary structure of a protein.   **1**
b) Describe the secondary structure of protein with reference to the histone 4 (H4) diagram.   **2**
c) The secondary structure of a protein can be altered more easily than the primary structure. Explain this in terms of the bonds involved.   **1**
d) (i) Give the term used to describe a component, such as the haem group shown in haemoglobin, embedded in a polypeptide.   **1**
   (ii) Explain the term 'co-operativity' in relation to oxygen binding to haemoglobin.   **1**

3 The phospholipid layer in membranes acts as a barrier to some molecules although others can pass through. Transmembrane proteins can act as channels or transporters to perform specific functions.

a) Describe the control of ligand-gated and voltage-gated channels.   **2**
b) Describe the role of ATPases in some active transport proteins.   **2**
c) The sodium–potassium pump is found in most animal cells and transports ions against a steep concentration gradient to create a membrane potential.
   (i) The mechanism of action of the sodium–potassium pump involves the stages shown below:
   - P – membrane protein is phosphorylated
   - Q – sodium ions bind to membrane protein
   - R – sodium ions are released
   - S – membrane protein changes conformation

   Give the correct sequence of stages in the action of the pump.   **1**
   (ii) Give **one** function of sodium–potassium pumps.   **1**

4 Insulin is a peptide hormone involved in the regulation of blood glucose in humans.

a) Describe how insulin is involved in the uptake of glucose into target cells.   **2**
b) Adiponectin is a signalling molecule thought to increase the sensitivity of cells to insulin.
   In a clinical study, the concentration of adiponectin in the blood of patients with type 2 diabetes was compared to non-diabetics. The results are shown in the table below.

Patient group	Average concentration of adiponectin in blood plasma ($\mu g\ cm^{-3}$)
Type 2 diabetic	6.6 ± 0.4
Non-diabetic	7.9 ± 0.5

   Explain how the results relate to the characteristics of type 2 diabetes.   **2**
c) Explain how regular exercise can contribute in reducing blood glucose concentration in subjects with type 2 diabetes.   **2**

5 In multicellular organisms, the process of apoptosis in target cells is triggered by cell death signals, which may originate within or outside the cell. The flowchart below shows the events triggered by a cell death signal from outside the cell.

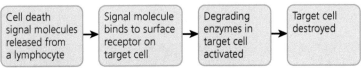

a) Give **one** reason why programmed cell death initiated by a lymphocyte can be beneficial to multicellular organisms.   1

b) (i) Name the degrading enzymes activated during apoptosis.   1

   (ii) Bcl-2 is a regulator protein that can inhibit apoptosis. In humans this protein is encoded by the BCL2 gene. Mutations in this gene can increase the levels of Bcl-2. Suggest why these mutations are usually associated with tumour growth.   2

c) (i) Give **one** example of an event originating within a cell that can trigger apoptosis.   1

   (ii) Name the tumour-suppressor protein activated by an internal death signal.   1

6 a) Give an account of the synthesis and transport of proteins under the following headings:

   (i) Synthesis of membrane components   6

   (ii) Movement of proteins for secretion   4

OR

b) Give an account of the process of photoreception in rod cells.   10

Answers are given on page 152

Area 2 Organisms and evolution

Key Area 2.1
Field techniques for biologists

Key points !

1 Aspects of fieldwork can present a **hazard**. ☐
2 Hazards in fieldwork include adverse weather conditions, difficult **terrain**, problems associated with isolation, and contact with harmful organisms. ☐
3 **Risk** is the likelihood of harm arising from exposure to a hazard. ☐
4 **Risk assessment** involves identifying risks and control measures to minimise them. ☐
5 Control measures to minimise risk in fieldwork include appropriate equipment, clothing and footwear, and means of communication. ☐
6 Sampling should be carried out in a manner that minimises impact on wild species and habitats. ☐
7 Consideration must be given to rare and vulnerable species and habitats that are protected by legislation. ☐
8 The chosen sampling technique, such as **point count**, **transect** or **remote detection**, must be appropriate to the species being sampled. ☐
9 A point count involves the observer recording all individuals seen from a fixed point count location. This can be compared to other point count locations or with data from the same location gathered at other times. ☐
10 **Quadrats**, of suitable size and shape, or transects are used for plants and other sessile or slow-moving organisms. ☐
11 Capture techniques, such as traps and nets, are used for mobile species. ☐
12 **Elusive species** can be sampled directly using camera traps or an indirect method, such as **scat sampling**. ☐
13 Identification of an organism in a sample can be made using classification guides, biological keys or analysis of DNA or protein. ☐
14 Organisms can be classified by both **taxonomy** and **phylogenetics**. ☐
15 Taxonomy involves the identification and naming of organisms, and their classification into groups based on shared characteristics. ☐
16 Classic taxonomy classification is based on morphology. ☐
17 Phylogenetics is the study of the evolutionary history and relationships among individuals or groups of organisms. ☐
18 Phylogenetics is changing the traditional classification of many organisms. ☐
19 Phylogenetics uses heritable traits such as morphology, DNA sequences and protein structure to make inferences about an organism's evolutionary history and create a phylogeny or phylogenetic tree – a diagrammatic hypothesis of its relationships to other organisms. ☐
20 Genetic evidence can reveal relatedness obscured by **divergent** or **convergent evolution**. ☐
21 Familiarity with taxonomic groupings allows predictions and inferences to be made about the biology of an organism from better-known **model organisms**. ☐
22 **Nematodes**, **arthropods** and **chordates** are examples of taxonomic groups. ☐
23 Model organisms are organisms that are either studied easily or have been well studied. ☐

⇨

⇨

24 Model organisms – for example, the bacterium *Escherichia coli*; the flowering plant *Arabidopsis thaliana*; the nematode *Caenorhabditis elegans*; the arthropod *Drosophila melanogaster* (a fruit fly); mice, rats and zebrafish, which are all chordates – have been very important in the advancement of modern biology. ☐

25 Information obtained from model organisms can be applied to other species that are more difficult to study directly. ☐

26 The presence, absence or abundance of **indicator species** can provide information about environmental qualities, such as the presence of a pollutant. ☐

27 Susceptible and favoured species can be used to monitor an ecosystem. ☐

28 The absence of an indicator species or reduced population indicates it is susceptible to some factor in the environment; its abundance or increased population indicates it is favoured by the conditions. ☐

29 The **mark and recapture** technique is used to estimate population size using the formula N = MC / R. ☐

30 A sample of the population is captured and marked (M) and released. After an interval of time, a second sample is captured (C). If some of the individuals in this second sample are recaptured (R), then the total population (N) can be estimated. ☐

31 The mark and recapture technique assumes that all individuals have an equal chance of capture, that there is no immigration or emigration, and that individuals that are marked and released can mix fully and randomly with the total population. ☐

32 Methods of marking animals include banding, tagging, surgical implantation, painting and hair clipping. ☐

33 The method of marking and subsequent observation must minimise the impact on the study species. ☐

34 Some of the measurements used to quantify animal behaviour are **latency, frequency** and **duration**. ☐

35 Latency is the time between the stimulus occurring and the response behaviour; frequency is the number of times a behaviour occurs within the observation period; duration is the length of time each behaviour occurs during the observation period. ☐

36 An **ethogram** is a list of species-specific behaviours shown by a species in a wild context that allows the construction of **time budgets**. ☐

37 Recording the duration of each of the behaviours in an ethogram, together with the total time of observation, allows the proportion of time spent on each behaviour to be calculated as a time budget. ☐

38 **Anthropomorphism** is the attribution of human characteristics or behaviour and emotions to an animal's behaviour. ☐

39 It is important to avoid anthropomorphism when analysing behaviour because it can lead to invalid conclusions. ☐

Fieldwork in biology

Fieldwork is a research activity done in natural habitats outside the laboratory. Biological samples obtained during fieldwork may later be processed in a laboratory, of course. In laboratory experiments, it is relatively easy to control variables whereas in the field, variables are extremely difficult and often impossible to control.

Aspects of fieldwork can present hazards, including adverse weather conditions; difficult terrain such as cliffs, marshes and intertidal areas; problems associated with isolation; and contact with harmful organisms. Risk is the likelihood of harm arising from exposure to any of these hazards; risk assessment involves identifying and evaluating these hazards, and identifying control measures needed to minimise the risk. These measures include appropriate equipment, clothing, footwear, and means of communication.

Equipment

The appropriate equipment is essential. Fieldworkers should ensure that any vehicles being used are serviced and fuelled. They should also carry a first aid kit with any specialist drugs related to the habitats in which they are working, such as specific insect repellents. Field equipment should be light and portable, with charged batteries if required. Recording sheets, either paper or digital, will be essential, and appropriate containers for samples and specimens will be needed. Optical equipment such as lenses, binoculars, telescopes and cameras should be clean and in working order.

Clothing and footwear

Appropriate clothing might include windproof and waterproof outer garments as well as specialist clothing such as gauntlets for handling venomous species or veils and hoods for swarming insects. Appropriate footwear might include boots, wellingtons, waders or gaiters in areas where insects such as ticks are common.

Communication and orientation

It is essential that fieldworkers are in communication with the outside world, especially if they are working in isolated areas. Weather forecast should be consulted before starting out. Maps, appropriate permits, a compass, fully charged mobiles with spare external batteries and walkie-talkies are essential. Information on routes and timescale should be left with the appropriate authorities, and team leaders should carry home contact details for all members of their team and for appropriate authorities. A formal risk assessment should be carried out – for an example of the type of form to be used see Figure 1.2 on page 9 – before fieldwork commences.

> **Check-up 25**
>
> 1 Explain the difference between a hazard and a risk.   **2**
> 2 Give **five** examples of health and safety precautions that should be taken on a field study being
>   carried out at high altitude in the mountains of Scotland.   **5**

Sampling

Sampling involves the capture and recording of wild species, or the measurement and recording of abiotic factors such as temperature or light intensity. Sampling should give data that is truly representative of the species or factors being sampled. It should always be carried out in a manner that minimises impact on wild species and their habitats. Special consideration must be given to rare and vulnerable species and habitats, which are protected by legislation that is legally binding and enforceable in law.

> **Key links** 👍
>
> There is much more about sampling in Key Area 3.2.

Minimising impacts

In some studies, wild species may be sampled using observation only. This is preferable but, in some cases, capture may be needed. If this is the case, consideration should be given to minimising the numbers needed to be caught and how sampled individuals can be returned to the wild as rapidly as possible and with minimum handling.

> **Hints & tips** ⭐
>
> *Make sure you read about the different sampling strategies in Key Area 3.2, not limiting yourself to knowing about the methods of sampling and being clear about what is meant by representative sampling.*

The mere presence of people in certain habitats can have a major impact. Trampling of vegetation; disturbance of non-target species, especially if breeding is underway; and the potential to unwittingly guide predators to prey species or their young must be taken seriously.

Techniques

The techniques used – such as point counts, transects or remote detection – must be appropriate to the species being sampled. A point count involves the observer recording all individuals seen from a fixed location, which can be compared to counts from other locations or with data from the same location gathered at other times. This technique is often used when investigating populations and distributions of large, diurnal species such as birds. Cameras, binoculars and telescopes are often used with this technique.

Transects are sampling lines laid across habitats affected by environmental gradients such as altitude, light intensity or tidal movements. Samples are collected at intervals called stations along the line. A line transect, as shown in Figure 2.1a), is normally used to study the distribution of a single plant species and those individuals touching the line at a station are counted in the sample. A belt transect samples a wider zone along a transect and can be used to study a community of plants or sessile animals. This involves the use of quadrats at each station, as shown in Figure 2.1b).

a) line transect

b) belt transect

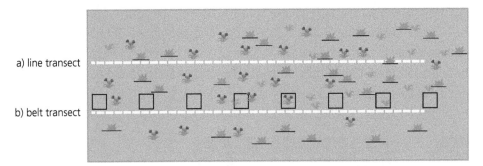

Figure 2.1 Comparison between a) a line transect and b) a belt transect

Quadrats

Quadrats come in many forms and sizes. Frame quadrats are simple square frames of various sizes, sometimes subdivided into smaller squares used for estimating the percentage ground cover of species, as shown in Figure 2.2a). Point quadrats have metal spikes that are pushed down into the plant community, as shown in Figure 2.2b); individuals touching the spike points are counted in.

a)

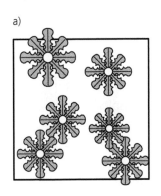

b)

Figure 2.2 a) a 0.5 m × 0.5 m frame quadrat containing six dandelion plants – the quadrat is 0.25 m² so there would be an estimated density of 24 dandelions per m²; percentage cover could also be estimated in this way, especially if the quadrat were divided into a grid; b) a point quadrat with ten pins – three pins have hit dandelions, so the frequency of dandelion occurrence is three out of ten, or 30%

Mobile species

Animals are usually mobile so have to be captured to identify them. Live trapping is used to do this. Traps take many forms, from simple pitfalls for ground invertebrates, as shown in Figure 2.3a), to mist nets for small birds, as shown in Figure 2.3b).

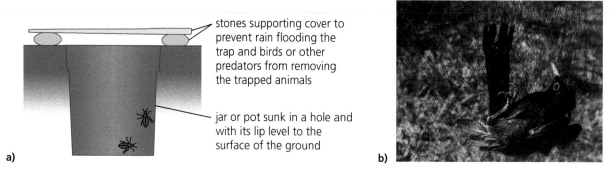

a)
stones supporting cover to prevent rain flooding the trap and birds or other predators from removing the trapped animals

jar or pot sunk in a hole and with its lip level to the surface of the ground

b)

Figure 2.3 a) Diagram of a pitfall trap for ground invertebrates and b) a blackbird caught in a mist net; in each case the animals in the sample can be recorded then released unharmed back into the habitat

Elusive species

Some species are elusive. They may be very rare, nocturnal, live in densely vegetated habitats or forest canopy, or have behaviour that makes them difficult to see. The presence of these species may be revealed using camera traps, as shown in Figure 2.4a), or by scat sampling. Scat is material such as faecal pellets that has been shed by the target animal. These can be treated using DNA analysis to confirm the species or even to identify an individual. Faeces can also be analysed to confirm their diet.

a)

b)

Figure 2.4 a) Camera trap for remote detection of elusive species and b) an extremely elusive species – the snow leopard

Check-up 26 ?

1 Explain what is meant by protective legislation.   **1**
2 Describe **two** precautions that can minimise the impact on species being sampled in the field.   **2**
3 Explain the differences between a line transect and a belt transect.   **2**
4 Describe a frame quadrat and the type of species it could be used for.   **2**
5 Describe what is meant by an elusive species and how its presence might be detected and monitored.   **3**

Identification and taxonomy

Identification of the species in a sample can be made using classification guides, dichotomous biological keys, or analysis of DNA or protein. Classification guides usually deal with a related group of species, such as birds, dragonflies or butterflies. They are often illustrated and give identification details as well as information on distribution, diet, migration and breeding times. Figure 2.5 shows some typical field guides.

Dichotomous keys can be used to identify more difficult groups and are usually laid out as a series of questions with answers that lead the user further into the key and, finally, to the identification. A simple example for some arthropods commonly found in pitfall traps is shown below.

Figure 2.5 Typical classification field guides – these ones are for birds

1 three pairs of legs .... go to 3
   more than three pairs of legs .... go to 2
2 four pairs of legs .... arachnid
   more than four pairs of legs .... go to 4
3 two pairs of large scaly wings .... lepidopteran
   one pair of wings .... dipteran
4 seven pairs of legs .... isopod
   more than seven pairs of legs .... go to 5
5 one pair of legs per segment .... chilopod
   two pairs of legs per segment .... diplopod

Organisms can be classified by both taxonomy and phylogenetics. Taxonomy involves the identification and naming of organisms, and their classification into groups based on shared characteristics. Classic taxonomy classification is based on morphology.

Phylogenetics

Phylogenetics is the study of the evolutionary history and relationships among individuals or groups of species, and it is changing the traditional classification of many species. Phylogenetics uses heritable traits, such as morphology, DNA sequences and protein structure, to make inferences about an organism's evolutionary history and to create a phylogeny or phylogenetic tree, which is a diagrammatic hypothesis of its relationships to other species.

Figure 2.6a) is a phylogenetic tree of some species of carnivorous mammals showing their evolutionary relatedness. Genetic evidence in DNA sequences can reveal relatedness that may have been obscured by divergent or convergent evolutionary processes. These processes can lead to confusion about how closely species are related, as shown in Figure 2.6b).

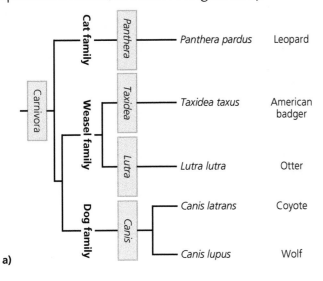

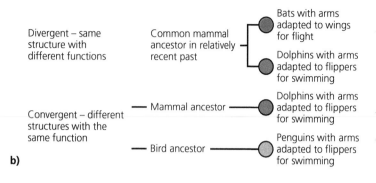

Divergent – same structure with different functions

Common mammal ancestor in relatively recent past

Bats with arms adapted to wings for flight

Dolphins with arms adapted to flippers for swimming

These look different but are closely related

Convergent – different structures with the same function

Mammal ancestor — Dolphins with arms adapted to flippers for swimming

Bird ancestor — Penguins with arms adapted to flippers for swimming

These look similar but are more distantly related

b)

Figure 2.6 a) Phylogenetic tree showing evolutionary relatedness based on DNA sequences; b) divergent and convergent processes can make evolutionary relatedness based on appearance a risky business

Convergence can lead to the assumption that species are closely related through their appearance when, in fact, they are only similar because evolutionary processes modified different structures for a similar function. Divergence can lead to the assumption that two species are very distantly related when, in fact, they are only different because the same structure has become adapted for different functions.

Example

In modern biology, new species are often discovered by DNA evidence suggesting that what had been regarded as a single species is actually two or more different ones. For example, the subalpine warbler, a European bird, has recently been split into three separate species – the Western subalpine warbler (*Sylvia inornata*), Moltoni's warbler (*Sylvia subalpina*) and the Eastern subalpine warbler (*Sylvia cantillans*) – based on DNA analysis.

Figure 2.7 Male Subalpine warbler *Sylvia cantillans*

Classification

The animal kingdom is divided into phyla, which includes:

- Arthropoda, made up of invertebrates with jointed legs and a segmented body, typically with paired appendages
- Nematoda, made up of round worms, which are very diverse and many of them are parasitic
- Chordata, made up of the vertebrates and others with a dorsal or spinal notochord.

Key links

There is more about parasites in Key Area 2.5.

Check-up 27

1	State the purpose of a dichotomous key.	2
2	Describe what is meant by taxonomy.	1
3	Explain what is meant by phylogeny and how it differs from taxonomy.	2
4	Explain the differences between divergent and convergent evolution.	2
5	Describe the main features of the phyla Arthropoda, Nematoda and Chordata.	3

Model species

Familiarity with taxonomic groupings allows predictions and inferences to be made between the biology of a species of interest and better-known model species or type species. Model species are chosen because they are common and usually easy to keep and study in captivity. Model species are used to obtain information that can be applied to species that are more difficult to study directly. They have been very important in the advancement of modern biology and include the bacterium *Escherichia coli*; the flowering plant *Arabidopsis thaliana*; the nematode *Caenorhabditis elegans*; the arthropod *Drosophila melanogaster* and chordates such as mice, rats and zebrafish.

Indicator species

Indicator species are those whose presence, absence or abundance can give information on environmental qualities, such as the presence of pollutant – for example, sulfur dioxide in the air or human faeces in water – or a desired human resource, such as copper or oil. Some plant indicators are used to verify the classification of a habitat. Each species exists in a niche, so their presence or absence can indicate if their niche requirements are present or not, and so may indicate information about abiotic factors such as temperature, oxygen, pH, salinity or mineral levels. Absence of a species or a reduced population indicates that the species is susceptible to some factor in the environment, while abundance or increased population indicates it is favoured by the conditions.

The table below lists some different indicator species and what they can indicate.

Taxonomic group	Type of organism	Indicator notes
Usnea species	Lichen	Absence indicates that sulfur dioxide pollution is present in the air
Chironomid larvae	Midge larvae	Abundance indicates a lack of dissolved oxygen in polluted fresh water
Erica cinerea	Bell heather	Presence indicates that a community could be classified as a gorse heath in Britain

Monitoring populations

The general condition of the environment can be checked using indicator species, but populations of animals and plants are monitored to check if they are increasing or decreasing. The Lincoln–Petersen method can be used to estimate animal population size if two visits are made to the study area. This method assumes that the study population is closed, which means that the two visits are near enough in time so that no individuals die, are born, move into the study area (immigrate) or move out of the study area (emigrate) between visits. The index involves marking animals, and also assumes that the marks do not disappear between visits to the site, and that the researcher correctly records all marks. Given those conditions, the population size is estimated using the following equation:

$$N = \frac{MC}{R}$$

Where N = estimated number of animals in the population, M = number of animals marked on the first visit, C = number of animals captured on the second visit, and R = number of recaptured animals that were marked.

> ### Hints & tips ⭐
>
> *If R is too low due to errors in marking, the population will be overestimated; if R is too high due to inadequate dispersal from the release site, the population will be underestimated.*

Marking methods

Marking methods used should minimise the impact on the survival of the study species and be appropriate to the purpose of marking. Banding is the placing of a visible coded loop or ring around a leg and is often used for birds as shown in Figure 2.8.

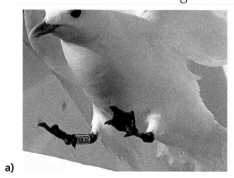
a)

b)

Figure 2.8 a) An ivory gull being released from capture – note that it has been banded with a yellow plastic ring carrying a serial number that will allow this individual to be identified using a telescope without recapture; it is durable enough to last the lifetime of the gull; it is also very light, so that flight is not affected and it is unlikely to make the gull an easier target for predators; b) a yellow wagtail of the Iberian race with an aluminium ring, which is more difficult to read in the field; it would need to be recaptured to be sure of its details

Tagging is similar and consists of a visible or GPS coded attachment. Surgical implantation involves inserting a microchip into the animal's tissues. Painting a code or mark on the outside of an animal is appropriate for an animal such as a snail or beetle. Hair clipping or ear punching involves making a unique hair clip or ear punch and is often appropriate for mammals.

Animal behaviour

Animal behaviour can be studied in wild populations using careful observation and measurement. Measurements might include:

- latency, which is the time between a stimulus occurring and the response behaviour being observed
- frequency, which is the number of times a behaviour occurs within the observation period
- duration, which is the length of time each behaviour occurs during the observation period.

An ethogram is a list or graphic display of the behaviours shown by a species in a wild context. It allows investigation and the construction of time budgets. An example of an ethogram is given in Figure 2.9.

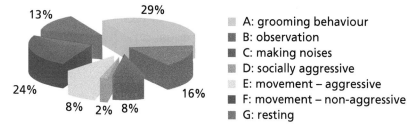

- **A: grooming behaviour**
- **B: observation**
- **C: making noises**
- **D: socially aggressive**
- **E: movement – aggressive**
- **F: movement – non-aggressive**
- **G: resting**

Figure 2.9 Behaviour of a chimpanzee during a study period showing an ethogram as a list of behaviour categories and a graphic of the time budget for the period

When constructing an ethogram it is important to avoid anthropomorphism, which is the allocation of human characteristics and emotions to the behaviour of animals. If this happens there is a risk that behaviour could be misinterpreted or that any conclusions reached could be invalid.

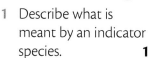

Check-up 28

1 Describe what is meant by an indicator species. **1**
2 Explain the terms latency, frequency and duration as measurements used to quantify animal behaviour. **3**
3 Explain how an ethogram of the behaviours shown by a species in a wild context allows the construction of time budgets. **2**
4 Explain anthropomorphism and the importance of avoiding it when analysing behaviour. **2**

Key words

Anthropomorphism – the attribution of human characteristics or behaviour and emotions to an animal's behaviour

Arthropods – joint-legged invertebrates that are identified by their segmented body, typically with pared appendages, for example *Drosophila melanogaster* (a fruit fly), wasps, butterflies, spiders and crabs

Chordates – sea squirts and vertebrates, for example birds, mammals, reptiles, amphibians and fish

Convergent evolution – the independent evolution of similar features in different species

Divergent evolution – the process whereby groups from the same common ancestor evolve and accumulate differences, resulting in the formation of new species

Duration – the length of time each behaviour occurs during the observation period

Elusive species – species that are difficult to see because of their habitat, behaviour or rarity

Ethogram – lists species-specific behaviours to be observed and recorded in the study

Frequency – the number of times a behaviour occurs within the observation period

Hazard – any source of potential damage, harm or adverse health effects on an individual

Indicator species – species that, by their presence, absence or abundance, can give information about an environmental factor

Latency – the time between the stimulus occurring and the response behaviour

Mark and recapture – method for estimating population size; a sample of the population is captured, marked and released (M); after an interval of time, a second sample captured (C); if some individuals in the second sample are recaptures (R), estimate of the total population (N) can be calculated

Model organisms – organisms that are either easily studied or have been well studied to provide information that can be applied to other species that are more difficult to study directly

Nematodes – also called roundworms; unsegmented thread-like body; many of them are parasitic, living inside their host, for example *Caenorhabditis elegans*

Phylogenetics – the study of the evolutionary history and relationships among individuals or groups of organisms

Point count – sampling technique that involves the observer recording all individuals seen from a fixed location

Quadrat – square frame of known area for sampling sessile organisms along a belt transect

Remote detection – ability to shown an animal's presence from a distance using a camera trap or scat sampling

Risk – the likelihood of harm arising from exposure to a hazard

Risk assessment – involves identifying risks and control measures to minimise them

Scat sampling – sampling technique used for elusive species whereby animal droppings are collected, which provide information about species abundance and diet

Taxonomy – the identification and naming of organisms, and their classification into groups, based on shared characteristics

Terrain – physical geography of the land

Time budget – the amount or proportion of time that animals spend in different behaviours, or in performing different classes of behaviour

Transect – a line or belt across a habitat or part of a habitat along which the number of organisms of each species can be observed and recorded at regularly placed stations

Exam-style questions

Structured questions

1 *Prestonella bowkeri* is a small, terrestrial snail found in rocky cliff face habitats on the Great Escarpment of southern Africa. The species lives in cracks and crevices on cliffs where it can be reliably located and is easy to catch.

In an attempt to estimate how its population changes at one small cliff site, samples of snails were captured, marked and released on two separate occasions during a year of study, as shown in the table below.

Date of sampling	Number of snails captured and marked during sampling (M)	Number of snails in second sample (C)	Number of marked snails in second sample (R)	Population estimate (N)
March	1435	1725	195	
September	1400		250	7000

  a) Use the formula $N = \dfrac{MC}{R}$ to:

    (i) calculate the population estimate for the March sample   **1**

    (ii) calculate the number of the snails in the second September sample.   **1**

  b) (i) Describe an appropriate sampling method for a slow-moving mollusc with predictable behaviour at a small site such as this.   **1**

    (ii) Suggest an appropriate method for marking individual snails, and describe factors that should be considered when choosing the method to be used.   **2**

2 Investigators observed behaviour of a social group of 15 chimpanzees in the wild. They devised a checklist of different actions which they then used in the analysis of the behaviour observed.

  a) Give the name used for a behavioural checklist of this kind.   **1**

  b) The investigators noted times at which different behaviours started and finished for different individuals and used these to calculate aspects of behaviour such as latency.

State what is meant by latency in animal behaviour studies.   **1**

  c) In devising the checklist, some statements included words such a 'grinning' and 'annoyed gestures'. State the term used for using human emotions in an ethogram and explain why it should be avoided in behavioural studies.   **2**

  d) The investigators used direct observation through binoculars to monitor the animals' behaviour but were concerned that some details might have been missed.

Suggest an improvement to their method that would reduce this source of error.   **1**

Extended response

3 Write notes on the different methods that can be used to sample wild populations.   **5**

4 Discuss the problems of evolutionary divergence and convergence to the study of phylogenetic relatedness in animal groups.   **4**

5 Write notes on the importance of the following categories of organisms in biology:

  a) model species   **3**

  b) indicator species   **3**

Answers are given on pages 153–154.

Key Area 2.2
Evolution

Key points !

1 **Evolution** is the change over time in the proportion of individuals in a population differing in one or more inherited traits. ☐

2 During evolution, changes in allele frequency occur through the non-random processes of **natural selection** and **sexual selection**, and the random process of **genetic drift**. ☐

3 Natural selection acts on genetic variation in populations. ☐

4 Variation in traits arises as a result of mutation. ☐

5 Mutation is the original source of new sequences of DNA, and these sequences can be novel alleles. ☐

6 Most mutations are deleterious or neutral, but in rare cases they may be advantageous to the fitness of an individual. ☐

7 Populations produce more offspring than the environment can support. ☐

8 Individuals with variations that are better suited to their environment tend to survive longer and produce more offspring, breeding to pass on those alleles that conferred an advantage to the next generation. ☐

9 Selection results in the non-random increase in the frequency of advantageous alleles and the non-random decrease in the frequency of deleterious alleles. ☐

10 Sexual selection is the non-random process involving the selection of alleles that increase the individual's chances of mating and producing offspring. ☐

11 Sexual selection may lead to differences in form between male and female individuals known as **sexual dimorphism**. ☐

12 Sexual selection can be due to **male–male rivalry** and **female choice**. ☐

13 Male–male rivalry: large size or weaponry increases access to females through success in conflict. ☐

14 Female choice involves females assessing the **fitness** of males. ☐

15 Genetic drift occurs when chance events cause unpredictable fluctuations in allele frequencies from one generation to the next. ☐

16 Genetic drift is more important in small populations, as alleles are more likely to be lost from the **gene pool**. ☐

17 The population **bottleneck effect** is an example of genetic drift that can occur when a population size is reduced for at least one generation. ☐

18 **Founder effects** are examples of genetic drift that occur through the isolation of a few members of a population from a larger population and the gene pool of the new population is not representative of that in the original gene pool. ☐

19 A gene pool is altered by genetic drift because certain alleles may be under-represented or over-represented, and so allele frequencies change. ☐

20 **Selection pressures** are the environmental factors that influence which individuals in a population pass on their alleles. ☐

21 Selection pressures can be biotic (competition, predation, disease, parasitism) or abiotic (temperature, light, humidity, pH, salinity). ☐

22 Where selection pressures are strong, the rate of evolution can be rapid. ☐

23 The **Hardy–Weinberg (HW) principle** states that, in the absence of evolutionary influences, allele and genotype frequencies in a population will remain constant over the generations. ☐

24 The conditions for maintaining the HW equilibrium are: no natural selection, random mating, no mutation, large population size and no gene flow through migration, in or out. ☐

25 The HW equation can be used to determine whether a change in allele frequency is occurring in a population over time, which would suggest that evolutionary process are occurring. ☐

⇨

⇨

26 The HW equation is: $p^2 + 2pq + q^2 = 1$, where p^2 = frequency of homozygous dominant genotype, $2pq$ = frequency of heterozygous genotype and q^2 = frequence of homezygous recessive genotype. ☐

27 Fitness is an indication of an individual's ability to be successful at surviving and reproducing. ☐

28 Fitness is a measure of the tendency of some organisms to produce more surviving offspring than competing members of the same species. ☐

29 Fitness refers to the contribution made to the gene pool of the next generation by individual genotypes. ☐

30 **Absolute fitness** is the ratio between the frequency of individuals of a particular genotype after selection, to those before selection. ☐

31 If the absolute fitness is 1, then the frequency of that genotype is stable. A value greater than 1 conveys an increase in the genotype, while a value less than 1 conveys a decrease. ☐

32 **Relative fitness** is the ratio of the number of surviving offspring per individual of a particular genotype to the number of surviving offspring per individual of the most successful genotype. ☐

33 **Co-evolution** is the process by which two or more species evolve in response to selection pressures imposed by each other. ☐

34 A change in the traits of one species acts as a selection pressure on the other species. ☐

35 Co-evolution is frequently seen in pairs of species that have symbiotic interactions. ☐

36 **Symbiosis** is a co-evolved intimate relationship between members of two different species. ☐

37 The impacts of symbiotic relationships can be positive (+), negative (−) or neutral (0) for the individual species involved. ☐

38 **Mutualism** is a symbiosis in which the species in the interaction are interdependent on each other for resources or other services and, since both species gain, the interaction is (+/+). ☐

39 **Commensalism** is a symbiosis in which only one of the species substantially benefits and, for the other, the relationship is neither substantially positive or negative, so is neutral (+/0). ☐

40 **Parasitism** is a symbiosis in which the parasite species benefits in terms of energy or nutrients while the host species is harmed by the loss of these resources (+/−). ☐

41 The **Red Queen hypothesis** states that, in a co-evolutionary relationship, change in the traits of one species can act as a selection pressure on the other species, which must adapt to avoid extinction. ☐

Introducing evolution

Evolution is the change over time in the proportion of individuals in a population differing in one or more inherited traits. Evolution can occur through the random process of genetic drift or the non-random processes of natural selection related to resources for survival and sexual selection related to reproduction.

Genetic drift

Genetic drift occurs when chance events cause unpredictable fluctuations in allele frequencies from one generation to the next. Genetic drift is more important in small populations as alleles are more likely to be completely lost by chance from a small gene pool, and any change in allele frequency is likely to be more significant to the population as a whole. Certain alleles may be randomly under-represented or over-represented following drift since the new population carries only a

> **Hints & tips** ⭐
>
> *Be sure to realise that the material in this Key Area is more detailed and complex than the ideas about evolution you studied at Higher or N5 levels.*

random sample of the original population's alleles. Two commonly occurring examples of the effects of genetic drift are recognised:

● Population bottlenecks occur when a population size is randomly reduced for at least one generation, so lowering the range of alleles upon which any subsequent selection pressure may then act.

● Founder effects occur through the isolation of a few random members of a population from a larger population, so the gene pool of the new population is not representative of that in the original gene pool.

Figure 2.10 shows a model of how genetic drift works.

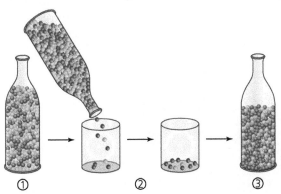

① ② ③

Figure 2.10 ① The container represents the original population and the beads are its gene pool; ② a random sample of the population with a more limited gene pool survives a chance event, such as an earthquake, or becomes the founder population of a new colony; ③ the survivors or the founders breed but the resulting population has a changed gene pool; in this case the red bead allele is missing from the new population and the blue bead allele is more frequent

Selection

Selection acts on genetic variation in populations. Variation in traits arises as a result of mutation, which is the original source of new sequences of DNA. These new sequences can be novel alleles. Most mutations are harmful (deleterious) or neutral, but in rare cases they may be beneficial and give selective advantage and improve the fitness of an individual. Fitness refers to the overall ability to survive and breed. Selection results in the non-random increase in the frequency of advantageous alleles, which increase fitness, and the non-random decrease in the frequency of deleterious alleles, which reduce fitness.

Because populations tend to produce more offspring than the environment can support, there is competition for resources. Individuals with variations that are better suited to their environment tend to compete better, and so survive longer and produce more offspring than their less well-suited competitors. Survivors breed to pass on those alleles that conferred the advantage to the next generation, and so these alleles increase in frequency and become widespread very quickly.

Sexual selection is a non-random process involving the selection of alleles that increase an individual's chances of mating and producing offspring. These alleles don't necessarily help in the general survival of the individual in terms of obtaining food or avoiding predators. For this reason, sexual selection is thought to be more important in habitats where food is abundant and predation is low. Sexual selection may lead to the evolution of sexual dimorphism in which the appearance and behaviour of males and females become different, as shown in Figure 2.11.

Hints & tips

Remember DAN — Deleterious, selectively Advantageous and Neutral mutations.

Hints & tips

When answering Advanced Higher questions about natural selection, your answers are expected to be appropriately complex — use the terms 'selective advantage' and 'fitness' in your answers.

Sexual selection can occur through male–male rivalry or female choice. In male–male rivalry, large size or weaponry – such as antlers, tusks, jaws, teeth and horns – increase an individual's access to females through successful conflict with competing males, as shown by red deer rutting in Figure 2.12a).

Female choice involves females assessing the fitness of males through observation of honest signals related to display, colours, plumes, calls and songs, as shown by male black grouse lekking in Figure 2.12b).

Figure 2.11 Sexual dimorphism in mallards – the male above is brightly coloured with a green head and yellow bill, whereas the female (behind) is duller and camouflaged to be inconspicuous at its nest on the ground

Key links 👍

There is more about sexual selection, sexual dimorphism and honest signals in Key Area 2.4.

a)

b)

Figure 2.12 a) Male weaponry – the antlers of a red deer stag; b) female choice – male black grouse display so that their appearance and performance can be assessed by females

Hints & tips ⭐

Make sure you have the difference between male–male rivalry and female choice clear in your mind – it's slightly tricky.

Selection pressure

Selection pressure measures how strongly a biotic or abiotic factor in the environment exerts its effect by influencing which individuals in a population survive and pass on their alleles to the next generation.

Differences in selection pressure can influence the rate of evolution. Where selection pressures are strong, the rate of evolution can be rapid. Selection pressures can be biotic, including competition, predation, disease and parasitism, or abiotic, including changes in temperature, light, humidity, pH and salinity.

Check-up 29 ❓

1	Describe how genetic drift causes changes in allele frequency.	**2**
2	Describe the possible effects of mutation on fitness.	**3**
3	Explain what is meant by selection pressure.	**2**
4	Describe **two** methods through which sexual selection is achieved.	**2**

Hardy–Weinberg principle

The Hardy–Weinberg (HW) principle states that, in the absence of evolutionary influences, allele and genotype frequencies in a population will remain constant over the generations. The Hardy–Weinberg equilibrium can only apply under a set of conditions, including the absence of natural selection, a system of random mating, the absence of mutations and gene flow by migration, and a large population size. These conditions would not always be expected in natural situations, so changes to Hardy–Weinberg frequencies can indicate if evolutionary influences are active. Changes in allele frequency revealed by changes in Hardy–Weinberg equilibrium suggest evolution is occurring.

Hardy–Weinberg equation

Hardy–Weinberg is usually applied to a pair of alleles, p and q, where p is dominant and q is recessive. In a sexually reproducing population, the male gametes could be either p or q and the female gametes could be either p or q. In each case, the gametes would exist in equal numbers and every possible fertilisation would be equally likely to occur, as shown in the grid in Figure 2.13a). There are four possible fertilisations – p^2, q^2 – and two possible ways of obtaining pq, giving the Hardy–Weinberg equation shown in Figure 2.13b).

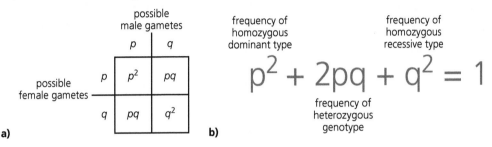

possible male gametes

	p	q
p	p^2	pq
q	pq	q^2

possible female gametes

a)

frequency of homozygous dominant type

frequency of homozygous recessive type

$$p^2 + 2pq + q^2 = 1$$

frequency of heterozygous genotype

b)

Figure 2.13 a) The Hardy–Weinberg principle; b) the Hardy–Weinberg equation, in which 1 represents the entire population – 100% of the individuals present

Example 🚩

The peppered moth, *Biston betularia*

In some parts of the UK, there are populations of this species with two different morphs. One has black or melanistic wings caused by the dominant allele B, and the other white or peppered wings caused by the recessive allele b, as shown in Figure 2.14. Wing colour is an important adaptation in these moths since they rely on camouflage to avoid predation when resting by day on exposed surfaces. Their predators are small birds, which use sight to detect their prey items.

a)

b)

Figure 2.14 The two morphs of *Biston betularia*; a) with melanistic wings and b) with peppered wings – wing colour is an adaptation to avoid predation

Suppose that studies showed that the frequency of the melanistic morph allele B in a population was 0.2, and the frequency of the population of the peppered morph allele b was 0.8. The Hardy–Weinberg equation would give:

$0.2^2 + 2 (0.2 \times 0.8) + 0.8^2 = 1$

Check-up 30 ❓

1	Give **five** conditions that must be met so that Hardy–Weinberg equilibrium can apply.	**5**
2	Give the Hardy–Weinberg equation.	**1**
3	Describe the conclusion that could be drawn if changes to Hardy–Weinberg equilibrium were detected.	**1**

⇨

Working this out and converting to phenotypes gives the following result:

$$\underbrace{(0.04 + 0.32)}_{\text{Melanistic 36\%}} + \underbrace{(0.64)}_{\text{Peppered 64\%}} = 1$$

In a population of 100 moths, the expected numbers of melanistic individuals would be 36 and the number of peppered individuals would be 64. If subsequent studies of this moth population showed that these numbers had changed, then this would suggest that the Hardy–Weinberg equilibrium had been broken and that evolutionary processes may be occurring.

Fitness

Fitness is an indication of an individual's success in surviving and reproducing. It is a measure of the tendency of some individuals to produce more surviving offspring than competing members of the same species and refers to the contribution made to the gene pool of the next generation by individual genotypes in one generation. It can be defined in absolute or relative terms.

Absolute fitness

Absolute fitness is taken as the ratio of the frequency of a particular genotype in a generation (G2) to its frequency in the previous generation (G1) as shown in Figure 2.14. If the absolute fitness is 1.0, then the frequency of that genotype is stable. A value greater than 1.0 conveys an increase in the genotype, and a value less than 1.0 conveys a decrease.

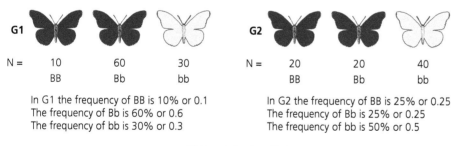

G1				G2			
N =	10	60	30	N =	20	20	40
	BB	Bb	bb		BB	Bb	bb

In G1 the frequency of BB is 10% or 0.1
The frequency of Bb is 60% or 0.6
The frequency of bb is 30% or 0.3

In G2 the frequency of BB is 25% or 0.25
The frequency of Bb is 25% or 0.25
The frequency of bb is 50% or 0.5

These results show that:
- Absolute fitness of BB is 0.25 : 0.1 = 2.5
- Absolute fitness of Bb is 0.25 : 0.6 = 0.4
- Absolute fitness of BB is 0.50 : 0.3 = 1.7

Figure 2.15 Absolute fitness: numbers (N) of genotypes in two generations (G1 and G2) of the peppered moth with calculations of absolute fitness

Relative fitness

Relative fitness is the ratio of the number of surviving offspring per individual of a particular genotype to the number of surviving offspring per individual of the most successful genotype, as shown in Figure 2.16.

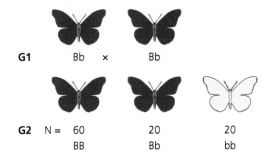

G1   Bb   ×   Bb

G2   N =   60   20   20
             BB   Bb   bb

The most successful genotype is BB, so the results show that:

- Relative fitness of BB is 60 / 60 = 1.0
- Relative fitness of Bb is 20 / 60 = 0.33
- Relative fitness of bb is 20 / 60 = 0.33

Figure 2.16 Relative fitness: numbers (N) of genotypes in the offspring (G2) of a pair of heterozygous peppered moths with calculations of relative fitness of the offspring genotypes

Hints & tips

Fitness is a tricky idea — make sure you realise that it describes the quality of an individual's adaptations, and so its chances of surviving and reproducing.

Check-up 31

1   Explain what is meant by fitness.   **2**
2   Describe the calculation needed to work out absolute fitness in a population.   **2**
3   Describe what is meant by relative fitness in a population.   **2**

Co-evolution

Co-evolution is the process by which two or more species evolve in response to selection pressures imposed by each other. A change in the traits of one species acts as a selection pressure on the other species. Co-evolution is frequently seen in pairs of species that have symbiotic interactions, which are intimate relationships between members of two different species. The impacts of these relationships can be positive (+), negative (−) or neutral (0) for the individual species involved.

Parasitism

In parasitism, the parasite has a positive (+) beneficial relationship with its host, which is negatively (−) affected by the relationship. The parasite gains habitat and resources and the host loses these resources.

Example

Plasmodium and malaria in humans

The parasitic protozoan *Plasmodium* causes malaria in humans. It can be introduced into human blood through the bite of an infected mosquito, as shown in Figure 2.17. The parasite gains a habitat, energy and resources from the humans, who are harmed by the loss of these and by the severe illness they suffer through the infection.

Figure 2.17 Mosquito biting a human host – the parasite *Plasmodium*, which causes malaria, can be transferred to humans in the bite from the mosquito, which acts as both the definitive host and a vector

Key links

There is a lot more about parasitism in Key Area 2.5.

Mutualism

In mutualism, the two species both derive positive (+) benefits and, in most cases, each species cannot survive without the other.

Clownfish and sea anemones

Clownfish live among the stinging tentacles of sea anemones, as shown in Figure 2.18. They protect the anemones from the predation by butterfish, which can eat the anemones, and will drive them away from the area. The anemone can also absorb nutrients from clownfish waste products. Clownfish gain because the anemone's stinging tentacles give them protection against their own predators and also provide a safe nest site in which to lay their eggs.

Figure 2.18 Mutualism between clownfish and anemones

Commensalism

In commensalism, one species derives positive (+) benefit and the other in the relationship is neither substantially positively nor negatively affected, so can be described as being neutral. In practice, it is a difficult symbiosis to show.

Water buffalo and cattle egrets

Cattle egrets are birds that eat large insects in wet grasslands. They often associate with large mammals such as water buffalo, as shown in Figure 2.19, walking between their legs and sometimes riding on their backs. The egrets gain from this because the buffalo disturb insects from the wet grass, while sitting up on the buffalo's back moves the egrets around and gives them a better view, all of which improves their ability to feed. The buffalo don't seem to substantially gain or lose from this behaviour.

Figure 2.19 Apparent commensalism in water buffalo and egrets – the buffalo disturb large insects from the wet grasslands, making it easy for the egrets to catch them for food; there seems to be little or no gain or loss for the water buffalo

Hints & tips

Remember that parasitism, mutualism and commensalism are all examples of symbiosis, and each species in the relationships is called a symbiont.

The Red Queen hypothesis

The Red Queen hypothesis states that in a co-evolutionary relationship, change in the traits of one species can act as a selection pressure on the other species, and that species must adapt to avoid extinction. Hosts that become better able to resist and tolerate their parasites have greater fitness. This puts selection pressure on these parasites to adapt and increase their fitness in terms of virulence through exploiting their hosts for food, reproduction and transmission to new hosts.

Key links

There is more about the Red Queen hypothesis in Key Area 2.3.

Check-up 32

1 Explain what is meant by symbiosis.   **3**
2 Use the +/−/0 notation to describe parasitism, commensalism and mutualism.   **3**
3 Describe the Red Queen hypothesis as it applies to parasites and their hosts.   **4**

Key words

Absolute fitness – the ratio between the number of individuals of a particular genotype after selection to those before selection

Bottleneck effect – a sharp reduction in the size of a population due to environmental events or human activities

Co-evolution – the process by which two or more species evolve in response to selection pressures imposed by each other

Commensalism – symbiosis in which only one of the species benefits (+/0)

Evolution – the change over time in the proportion of individuals in a population differing in one or more inherited traits

Female choice – a mechanism of sexual selection in which females assess males' fitness and choose the males with which they will mate

Fitness – a measure of the tendency of some organisms to produce more surviving offspring than competing members of the same species

Founder effects – when the gene pool of a new population is not representative of that in the original gene pool; occurs through the isolation of a few members of a population from a larger population

Gene pool – the total number of genes and their alleles in a population of one species

Genetic drift – the random increase or decrease in frequency of DNA sequences from one generation to the next, particularly in small populations

Hardy–Weinberg (HW) principle – in the absence of evolutionary influences, allele and genotype frequencies in a population will remain constant over the generations

Male–male rivalry – a mechanism of sexual selection in which males fight for females, often using weaponry such as antlers and tusks

Mutualism – both species in the interaction are interdependent on each other for resources or other services; as both organisms gain from the relationship, the interaction is (+/+)

Natural selection – the non-random increase in the frequency of advantageous alleles and the non-random decrease in the frequency of deleterious alleles

Parasitism – symbiosis in which the parasite benefits in terms of energy or nutrients and the host is harmed as the result of the loss of these resources (+/−)

Red Queen hypothesis – states that, in a co-evolutionary relationship, change in the traits of one species can act as a selection pressure on the other species

Relative fitness – the ratio of the number of surviving offspring per individual of a particular genotype to the number of surviving offspring per individual of the most successful genotype

Selection pressures – the environmental factors that influence which individuals in a population pass on their alleles   ⇨

Sexual dimorphism – females are generally inconspicuous; males usually have more conspicuous markings, structures and behaviours

Sexual selection – the non-random process involving the selection of alleles that increase the individual's chances of mating and producing offspring

Symbiosis – co-evolved intimate relationships between members of two different species

Exam-style questions

Structured questions

1 *Cyanea* is a genus of endemic flowering plants on the Hawaiian Islands. *Cyanea* is thought to have co-evolved with species of Hawaiian honeycreepers and honeyeaters, which serve as pollinators of their flowers. The birds visit flowers to obtain nectar from the bases of the flower tubes. As the birds probe the flowers with their beaks, pollen is brushed on to the feathering of their heads and can be carried to the next flower to achieve pollination. The diagram opposite shows the curvature and length of the flower tubes of two different *Cyanea* species and the heads of their main pollinators.

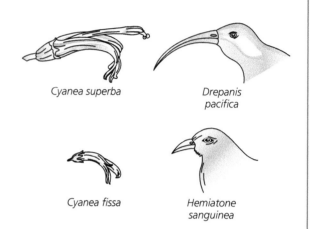

Cyanea superba

Drepanis pacifica

Cyanea fissa

Hemiatone sanguinea

   a) Explain what is meant by the term 'co-evolved'.   1
   b) Use the +/−/0 notation to describe the relationship between *Cyanea* and birds.   1
   c) Explain the advantage to *Drepanis pacifica* of its relationship with *Cyanea superba*.   2
   d) Describe how the Red Queen hypothesis can be used to explain the co-evolution of *Cyanea* plants and their bird pollinators.   3

2 a) In peppered moths, black wings (B) is dominant to peppered wings (b). In a population of these moths, 40% of the individual were peppered.
      Use the Hardy–Weinberg equation to calculate the following:
      (i)   the percentage of heterozygous moths in the population   2
      (ii) the frequency of homozygous dominant individuals.   2
   b) Give **three** conditions for the existence of Hardy–Weinberg equilibrium.   3

Extended response

3 Give an account of processes involved in evolution under the following headings:
   a) mutation   2
   b) genetic drift   3
   c) natural and sexual selection.   5
4 Write short notes on symbiosis under the following headings:
   a) parasitism   3
   b) mutualism and commensalism.   4

Answers are given on pages 154–155.

Key Area 2.3
Variation and sexual reproduction

Key points (!)

1 Sexual and asexual reproduction have both costs and benefits. ☐

2 Costs of sexual reproduction include males unable to produce offspring and only half of each parent's genome passed on to offspring, disrupting successful parental genomes. ☐

3 The benefits of sexual reproduction outweigh costs due to an increase in genetic variation in populations. ☐

4 Genetic variation provides the raw material required for adaptation, giving sexually reproducing species a better chance of survival under changing selection pressures. ☐

5 The Red Queen hypothesis can be used to explain the persistence of sexual reproduction. ☐

6 Co-evolutionary interactions between parasites and hosts may select for sexually reproducing hosts. ☐

7 Hosts better able to resist and tolerate parasitism have greater fitness; parasites better able to feed, reproduce and find new hosts have greater fitness. ☐

8 If hosts reproduce sexually, the genetic variability in their offspring reduces the chances that all will be susceptible to infection by parasites. ☐

9 Asexual reproduction can be a successful reproductive strategy as whole genomes are passed on from parent to offspring. ☐

10 In asexual reproduction, just one parent can produce offspring and establish a colony of virtually unlimited size over time. ☐

11 Maintaining the genome of the parent is an advantage, particularly in very narrow, stable niches or when recolonising disturbed habitats. ☐

12 Offspring can be reproduced more often and in larger numbers with asexual reproduction. ☐

13 **Vegetative cloning** in plants and parthenogenesis in lower plants and animals that lack fertilisation are examples of asexual reproduction in eukaryotes. ☐

14 **Parthenogenesis** is reproduction from a female gamete without fertilisation. ☐

15 Parthenogenesis is more common in cooler climates, which are disadvantageous to parasites, or regions of low parasite density or diversity. ☐

16 Asexually reproducing populations are not able to adapt easily to changes in their environment, but mutations can occur that provide some degree of variation and enable some natural selection and evolution to occur. ☐

17 Organisms that reproduce principally by asexual reproduction also often have mechanisms for **horizontal gene transfer** between individuals to increase variation, for example the plasmids of bacteria and yeasts. ☐

18 Prokaryotes can exchange genetic material horizontally, resulting in faster evolutionary change than in organisms that only use vertical transfer.

19 **Meiosis** is the division of the nucleus that results in the formation of haploid gametes from a diploid gametocyte. ☐

20 In diploid cells, chromosomes typically appear as **homologous pairs**. ☐

21 Homologous chromosomes are chromosomes of the same size, same centromere position and with the same sequence of genes at the same loci. ☐

22 The chromosomes, which have replicated prior to meiosis I, each consist of two genetically identical **sister chromatids** attached at the centromere. ☐

23 At meiosis I, chromosomes condense, the homologous chromosomes pair up and **chiasmata** form at points of contact between the non-sister chromatids of a homologous pair and sections of DNA are exchanged. ☐

⇨

⇨

24 **Linked genes** are those on the same chromosome and **crossing over** can result in new combinations of the alleles of these genes. ☐

25 Crossing over of DNA is random and produces genetically different chromosomes through **recombination**. ☐

26 Spindle fibres attach to the homologous pairs and line them up at the equator of the spindle. ☐

27 The orientation of the pairs of homologous chromosomes at the equator is random, and each pair of homologous chromosomes is positioned independently of the other pairs, irrespective of their maternal and paternal origin – this is **independent assortment**. ☐

28 The chromosomes of each homologous pair are separated and move towards opposite poles. ☐

29 **Cytokinesis** occurs and two daughter cells form. ☐

30 Each of the two cells produced in meiosis I undergoes a further division called meiosis II during which the sister chromatids of each chromosome are separated. ☐

31 A total of four haploid cells are produced as result of meiosis. ☐

32 The sex of birds, mammals and some insects is determined by the presence of **sex chromosomes**. ☐

33 In most mammals the **SRY gene** on the Y chromosome encodes testes-determining factor (TDF) and determines development of male characteristics. ☐

34 **Heterogametic** (XY) males lack most of the alleles corresponding to those on the X chromosome on their shorter Y chromosome. ☐

35 The presence of the short Y chromosome can result in sex-linked patterns of inheritance as seen with carrier females (X^BX^b) and affected males (X^bY). ☐

36 In **homogametic** females (XX), one of the two X chromosomes present in each cell is randomly inactivated at an early stage of development. ☐

37 **X chromosome inactivation** prevents a double dose of gene products, which could be harmful to cells. ☐

38 Carriers are less likely to be affected by deleterious mutations on one of their X chromosomes because of random inactivation. ☐

39 As the X chromosome inactivated in each cell is random, half of the cells in any tissue will have a working copy of the gene in question. ☐

40 **Hermaphrodites** are species that have functioning male and female reproductive organs in each individual. ☐

41 Hermaphrodites produce both male and female gametes, and usually need a partner with which to exchange gametes. ☐

42 The benefit to the individual hermaphrodite is that if the chance of encountering a partner is an uncommon event, there is no requirement for that partner to be of the opposite sex. ☐

43 For some species, environmental rather than genetic factors determine sex and sex ratio. ☐

44 Environmental sex determination in reptiles is controlled by environmental temperature of egg incubation. ☐

45 Sex can change within individuals of some species as a result of size, competition or parasitic infection. ☐

46 In some species the sex ratio of offspring can be adjusted in response to resource availability. ☐

Costs and benefits of sexual reproduction

Sexual reproduction is a widespread and common method of producing offspring that involves the combining of genetic information from two individuals, usually a male and a female of a species. Disadvantages of sexual reproduction include the fact that males, which generally make up half of the members of a species, are unable to produce offspring. Also, there can be disruption of successful parental genomes and only half of a parent's genome passes to their offspring.

However, the benefits of reproducing sexually seem to outweigh these disadvantages and include the increase in genetic variation in the population because of the mixing of parental genomes. This genetic variation provides the raw material required for adaptation, giving sexually reproducing organisms a better chance of survival under changing selection pressures.

The Red Queen hypothesis

The Red Queen hypothesis could explain the persistence of sexual reproduction in spite of its disadvantages. Co-evolutionary interactions between parasites and hosts may select for sexually reproducing hosts. Hosts better able to resist and tolerate parasitism have greater fitness. If hosts reproduce sexually, the genetic variability in their offspring reduces the chances that all will be susceptible to infection by parasites. However, parasites with increased virulence, better able to feed, reproduce and find new hosts also have greater fitness. If parasites reproduce sexually, the genetic variation in their offspring increases the chances of some of the offspring having improved ability to exploit their hosts. The selection pressure created by the evolved resistance of the hosts drives the evolution of the parasites to increase their virulence, as shown in Figure 2.20.

> **Hints & tips** ⭐
>
> *The Red Queen hypothesis is tricky to describe – make sure you use the terms 'selection pressure', 'virulence' and 'resistance'.*

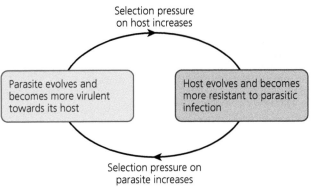

Figure 2.20 Red Queen hypothesis – note that the variation due to sexual reproduction is the material upon which natural selection works and allows the rates of evolution needed to support the hypothesis

> **Key links** 👍
>
> There is more about parasites in Key Areas 2.2 and 2.5.

Costs and benefits of asexual reproduction

Asexual reproduction involves the production of genetically identical offspring by a single parent. It can be a successful and very rapid reproductive strategy in narrow, stable niches or when recolonising disturbed habitats. These advantages arise because the offspring are genetically identical, which conserves advantageous genomes suited to a stable niche, and because offspring are developed rapidly, allowing early establishment in a newly created habitat.

Asexually reproducing populations are not able to adapt easily to changes in their habitat because their genetically identical offspring do not provide the variation required for natural selection. However, mutations, although rare, do arise and provide some variation and enable some natural selection and evolution.

In eukaryotes, examples of asexual reproduction include vegetative cloning in plants and parthenogenesis in animals, which involves the development of unfertilised eggs as shown in Figure 2.21. Parthenogenesis is more common in cooler climates that are disadvantageous to parasites or in regions of low parasite density and/or diversity.

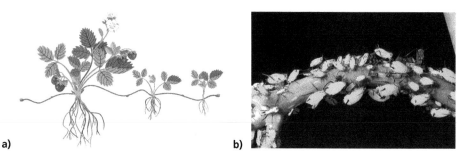

a)   b)

Figure 2.21 Asexual reproduction: a) strawberry plant clone and b) aphids, which can produce offspring by parthenogenesis

Horizontal gene transfer

Both sexual and asexual reproduction are examples of vertical gene transfer in which parent individuals of one generation pass copies of their genes on to offspring that form the next generation.

Organisms that reproduce principally by asexual reproduction often have mechanisms for horizontal gene transfer between individuals within a generation. Prokaryotes can exchange genetic material horizontally, such as in the transfer of plasmids between individual bacteria in a population, as shown in Figure 2.22.

the plasmid is taken up by another bacterial cell

bacterial cell releases plasmid into the environment

Figure 2.22 Horizontal gene transfer in a bacterial species – a copy of a plasmid containing an antibiotic resistance gene (blue region) is transferred from a donor to a recipient cell

Although yeast is eukaryotic, it is unusual in having plasmids that can also be transferred between individuals. Horizontal gene transfer can result in faster evolutionary change than in species that only use vertical transfer. For example, bacterial populations that are resistant to modern antibiotics are thought to have evolved very quickly due to increased exposure to antibiotics and the selection of resistance genes, present in plasmids, which were replicated and passed rapidly through bacterial populations.

Meiosis and variable gametes

Meiosis is a type of cell division that produces haploid gametes from diploid gamete mother cells. Homologous chromosomes are pairs of chromosomes in a genome that are of the same size, same centromere position and with the same genes at the same loci, as shown in Figure 2.23a). The variation in gametes produced depends on interactions and movements of homologous chromosomes.

Stages in meiosis

Meiosis can be thought of as two separate stages called meiosis I and meiosis II. Meiosis I involves pairing of homologous chromosomes followed by their separation into two cells. In meiosis II the sister chromatids are separated and each new chromosome segregates into one of the four haploid cells produced, as shown in Figure 2.24.

> ### Hints & tips
> Make sure you know that asexual reproduction is still vertical gene transfer – parent to offspring – even although some asexual reproducers can also inherit horizontally.

> ### Check-up 33
> 1 Describe the costs and benefits of sexual reproduction.   **3**
> 2 Explain how the Red Queen hypothesis can be used to explain the persistence of sexual reproduction.   **4**
> 3 Explain why asexual reproduction can be a successful reproductive strategy.   **2**
> 4 Describe the process of parthenogenesis and the conditions in which it is more common.   **3**
> 5 Explain what is meant by 'horizontal gene transfer' and its benefit to prokaryotes.   **3**

When homologous pairs separate during meiosis I, they do so independently and irrespective of their maternal and paternal origin.

This process of independent assortment leads to variation in the alleles reaching each gamete. The impact of these process is shown in Figure 2.23.

Homologous chromosomes undergo random crossing over at points called chiasmata, resulting in exchange of DNA and the recombination of alleles of linked genes, shown in Figure 2.23b), to produce varied combinations in the gametes, as shown in Figure 2.23c).

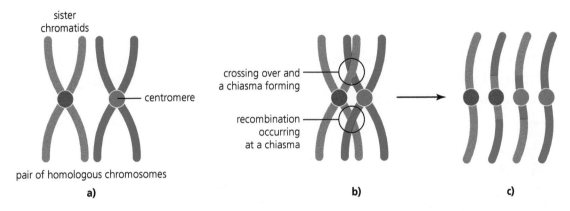

Figure 2.23 a) Pair of homologous chromosomes each made up from a pair of identical sister chromatids – each homologous chromosome has its centromere in the same position and the same genes at the same loci; b) crossing over occurring above centromere and recombination occurring below centromere; c) final daughter chromosomes showing variation from recombination

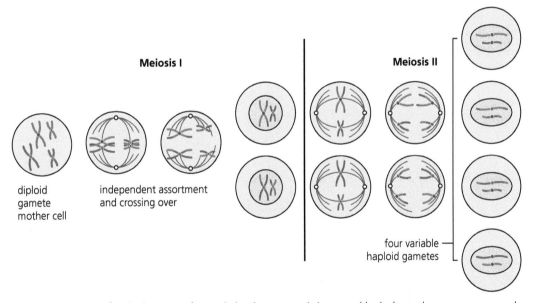

Figure 2.24 Stages of meiosis – note that variation in gametes is increased by independent assortment and crossing over shown during meiosis I

Check-up 34 ?

1 Describe homologous chromosomes.   2
2 Explain how chiasma formation between paired homologous
   chromosomes leads to variation in gametes.   2
3 Explain how independent assortment of chromosomes during meiosis I
   leads to variation in gametes.   2
4 Describe the main events during meiosis I.   5
5 Describe meiosis II.   2

Genetic sex determination

The sex of birds, mammals and some insects is determined by the presence of sex chromosomes. In mammals the two sex chromosomes are called X and Y, as shown in Figure 2.25a). Females have two X chromosomes (XX) and are said to be homogametic because all of their eggs contain an X chromosome. Males have an X and a Y (XY) and are said to be heterogametic because 50% of their sperm cells contain an X chromosome and the other 50% contain a Y. In most mammals a gene called SRY carried on the Y chromosome determines development of male characteristics by encoding and expressing a protein called testes-determining factor (TDF).

Because the Y chromosome is shorter than the X, heterogametic (XY) males have regions of their X chromosome that lack homologous alleles on their Y. This can result in sex-linked patterns of inheritance in which carrier females (X^BX^b) can pass recessive deleterious sex-linked alleles to give affected males (X^bY), as shown in Figure 2.25b).

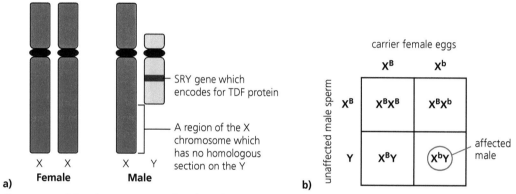

Figure 2.25 a) Sex chromosomes in female and male mammals; b) punnet square to show the possible offspring from a female carrier of a recessive sex-linked condition and an unaffected male

Hints & tips

You should be aware that males must express alleles on their single X chromosome because they have only one of these – if there is a deleterious allele, it will be expressed.

X chromosome inactivation

In homogametic females (XX) one of the two X chromosomes present in each cell is randomly inactivated at an early stage of development. X chromosome inactivation is a process by which most of one X chromosome is inactivated randomly in each cell. X chromosome inactivation prevents possible production of a double dose of gene products, which could be harmful to cells. As the X chromosome inactivated in each cell is random, half of the cells in any tissue will have a working copy of each sex-linked gene, as shown in Figure 2.26. This means that carriers of a sex-linked deleterious mutation are less likely to be affected by the mutation on one of their X chromosomes.

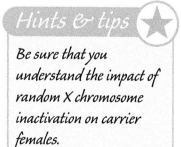

Hints & tips

Be sure that you understand the impact of random X chromosome inactivation on carrier females.

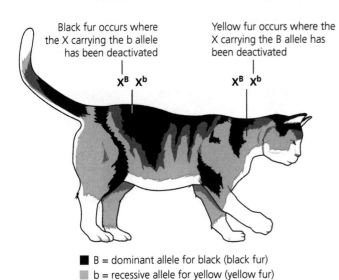

Black fur occurs where the X carrying the b allele has been deactivated

X^B X^b

Yellow fur occurs where the X carrying the B allele has been deactivated

X^B X^b

■ B = dominant allele for black (black fur)
▨ b = recessive allele for yellow (yellow fur)

Figure 2.26 Sex-linkage and X chromosome inactivation in the determination of coat colour in a female calico cat – female cats with the genotype X^BX^b have black and orange hairs in their coats depending on which of their X chromosomes has been deactivated by each cell. Why is it almost impossible for a calico cat to be male?

Check-up 35

1 Describe how genetic control determines the sex in mammals. **2**
2 Describe the function of the SRY gene. **1**
3 Explain why human males are said to be heterogametic. **1**
4 Explain the significance of heterogametic males in relation to sex-linked patterns of inheritance. **3**
5 Explain the importance of X chromosome inactivation and why female carriers of a sex-linked condition are less likely to be affected by any deleterious mutations on these X chromosomes. **3**

Environmental determination of sex and sex ratio

For some species, environmental rather than genetic factors determine sex and sex ratio, and sex can change within individuals of some species as a result of size, competition or parasitic infection. In some species, the sex ratio of offspring can be adjusted in response to resource availability.

Example

Body size in clownfish

Clownfish (see Figure 2.18 on page 80) live in groups with a large dominant female, a smaller subordinate male and some undifferentiated smaller subordinate juveniles. If the dominant female dies, subordinates grow in size. The male becomes female and the largest undifferentiated juvenile matures into a male. This ability allows the easy formation of a new breeding pair, preventing the need for dangerous travel across the habitat, but it requires the presence of undifferentiated fish in each group.

Competition for resources in some species of mouse lemur

Female mouse lemurs, as shown in Figure 2.27a), can detect the presence of a competing female lemur in their territory through chemicals in urine. The presence of a competing female can lead to production of a higher male to female ratio of offspring in her next litters. Males disperse away from their birth territory early compared with females, which tend to stay in their birth territory, and this has the advantage of reducing the competition for the mother lemur.

Parasitic infection by *Wolbachia* bacteria in some butterfly species

Wolbachia are bacteria that are passed into eggs by female butterflies, as shown in Figure 2.27b). They alter host biology by inducing feminisation in the offspring, which skews the sex ratio towards female. This subsequently favours the transmission of *Wolbachia* in eggs, which are only produced by females, of course.

Temperature in various reptile species

Environmental sex determination in reptiles is controlled by the environmental temperature of egg incubation. The sex ratio in offspring of Mississippi alligators, shown in Figure 2.27c), is temperature

⇨

dependent. Incubation at 31.5°C results in 100% female hatchlings, whereas an incubation temperature of 33°C results in 100% male hatchlings.

a)

b)

c)

Figure 2.27 a) Mouse lemur; b) butterfly with *Wolbachia*; c) Mississippi alligator

Hermaphrodites

Hermaphrodites are species that have functioning male and female reproductive organs in each individual, as shown by the molluscs in Figure 2.28. Each produces both male and female gametes and usually has a partner with which to exchange gametes. The benefit to the individual organism is that, if encountering a partner is an uncommon event, there is no requirement for that partner to be of the opposite sex – any individual will do.

Figure 2.28 Each hermaphrodite snail produces both male and female gametes

Check-up 36

1 Describe the environmental factors that determine sex and sex ratio in some species.   **4**
2 Explain why resource availability can change the sex ratio of offspring in some species.   **2**
3 Describe a hermaphrodite and explain the benefit to the individual organism of being hermaphrodite.   **2**

Key words

Chiasmata – a point at which paired chromosomes remain in contact during the first metaphase of meiosis, and at which crossing over and exchange of genetic material occurs between the strands

Crossing over – the exchange of genetic material between non-sister chromatids of two homologous chromosomes that results in recombinant chromosomes during meiosis

Cytokinesis – the physical process of cell division, which divides the cytoplasm of a parental cell into two daughter cells

Hermaphrodites – species that have functioning male and female reproductive organs in each individual

Heterogametic – dissimilar sex chromosomes, for example mammalian males where the Y chromosome is much smaller than the X chromosome, resulting in two kinds of gamete

Homogametic – sex chromosomes that do not differ in morphology, resulting in only one kind of gamete

Homologous pair – a pair of chromosomes of the same size, centromere position and sequence of gene; one is of maternal origin and the other paternal

Horizontal gene transfer – inheritance of genetic material within a generation

Independent assortment – formation of random combinations of chromosomes in meiosis and of genes on different pairs of homologous chromosomes by the passage of one of each diploid pair of homologous chromosomes into each gamete independently of each other pair

Linked genes – genes located on the same chromosome

Meiosis – the division of the nucleus that results in the formation of haploid gametes from a diploid gametocyte

Parthenogenesis – development of an offspring from a female gamete without fertilisation

Recombination – the establishment of new combinations of alleles following crossing over

Sex chromosomes – a pair of chromosomes that can determine sex in some species

Sister chromatids – the genetically identical strands of chromosomes

SRY gene – gene on the Y chromosome that determines development of a male by expressing a protein called testes-determining factor (TDF)

Vegetative cloning – any form of asexual reproduction occurring in plants in which a new plant grows from a fragment of the parent plant or a specialised reproductive structure

X chromosome inactivation – a process by which most of one X chromosome is randomly inactivated to prevent a double dose of gene products, which could be harmful to cells

Exam-style questions

Structured questions

1 **a)** The mechanism of sex determination differs between species.
    (i) In most mammals the sex of an individual is determined through its genotype. Describe how the genotype of a mammal can determine maleness.   **1**
    (ii) Give **two** examples of environmental factors that can determine sex or sex ratio in certain species.   **2**
   b) Haemophilia is a recessive sex-linked condition in humans determined by the allele X^b.
    (i) Explain why males are more likely to be affected by haemophilia than females.   **2**
    (ii) Explain why carrier females are less likely to be affected by recessive sex-linked conditions even though they have a copy of the faulty allele.   **1**
    (iii) A male with haemophilia and a female carrier are expecting a baby. If the baby is male, calculate the chances that the boy will be affected by haemophilia.   **1**

2 The diagram below shows the production of an offspring by vegetative cloning in a strawberry plant.

   a) This is an example of vertical gene transfer. Describe how this differs from horizontal gene transfer.   **2**
   b) Suggest **two** advantages to the strawberry plant of this type of asexual reproduction.   **2**
   c) Vegetative cloning is asexual and gives very little variability in offspring.

   (i) Explain how enough variation is produced to allow evolutionary processes such as natural selection to occur.   **2**
    (ii) Name the process of asexual reproduction in animals that involves the development of unfertilised eggs in haploid adult offspring.   **1**
    (iii) Give the term for an organism that produces both male and female gametes.   **1**

Extended response

3 Give an account of the production of variable gametes during meiosis.   **7**
4 Explain how the Red Queen hypothesis can be used to account for the persistence of sexual reproduction in many groups of animals despite the costs that it has.   **6**

Answers are given on pages 155–156.

Key Area 2.4
Sex and behaviour

Key points !

1 Sperm are produced in much larger numbers than eggs but each egg has a larger energy store. ☐

2 Females make large investments in the egg structure in non-mammals or in the uterus and during gestation in mammals. ☐

3 **Parental investment** is costly but increases the probability of production and survival of young. ☐

4 **r-selected** species are smaller, have a shorter generation time, mature more rapidly, reproduce earlier in their lifetime (often only once) and produce a larger number of smaller offspring, each of which receives a smaller energy input. They give limited parental care so most offspring will not reach adulthood. ☐

5 **K-selected** species are larger, live longer, mature more slowly, can reproduce many times in their lifetime and produce relatively few, larger offspring. They give a high level of parental care so many offspring have a high probability of surviving to adulthood. ☐

6 r-selection tends to occur in unstable environments where the species has not reached its reproductive capacity, whereas K-selection tends to occur in stable environments. ☐

7 Using external fertilisation allows very large numbers of offspring to be produced, but many gametes are predated or not fertilised, and there is no or limited parental care; few offspring survive. ☐

8 Using internal fertilisation increases the chance of successful fertilisation and fewer eggs are needed. Offspring can be retained internally for protection and/or development, and there is a higher offspring survival rate. ☐

9 In internal fertilisation, energy is used in locating a mate and it requires the difficult direct transfer of gametes from one partner to another. ☐

10 Mating systems are based on how many mates an individual has during one breeding season and range from **polygamy** in which individuals of one sex have more than one mate to **monogamy** in which individuals have only one mate. ☐

11 In **polygyny**, one male mates exclusively with a group of females and in **polyandry**, one female mates with a number of males in the same breeding season. ☐

12 Many animals have mate-selection courtship rituals, often involving **species-specific sign stimuli** and **fixed action pattern** responses in some birds and fish. ☐

13 Sign stimuli are actions or appearances given by individuals of a species that signal intent, often for courtship and mating. ☐

14 Fixed action patterns are instinctive sequences of behaviour in some species of birds and fish in which pieces of behaviour act as stimuli for responses, which in turn act as further stimuli. ☐

15 **Sexual selection** selects for characteristics that have little survival benefit for the individual, but increase their chances of mating. ☐

16 Many species exhibit **sexual dimorphism** as a product of sexual selection in which females are generally inconspicuous and males usually have more conspicuous markings, structures and behaviours. ☐

17 **Reversed sexual dimorphism** occurs in some species in which females are more conspicuous than males. ☐

18 Female choice involves females assessing **honest signals** of the fitness of males. ☐

19 Honest signals can indicate favourable alleles that increase the chances of survival of offspring (fitness) or a low parasite burden, suggesting a healthy individual. ☐

20 In lekking species, males gather to display at a **lek**, where female choice occurs. ☐

⇨

⇨

21 Dominant males occupy the centre of a lek, with subordinates and juveniles at the fringes as 'satellite' males. Female choice occurs during the display. ☐

22 Success in **male–male rivalry**, through real or ritualised conflict, increases access to females for mating. ☐

23 Males will fight for dominance and access to females, often using elaborate weaponry such as antlers, tusks and horns. ☐

Parental investment

In sexual reproduction, greater investment is made by females in the egg structure in non-mammals or in the uterus and during gestation in mammals, compared to the investment of males. Parental investment is costly but it increases the probability of production and survival of young.

Strategies for sexual reproduction can be classified into r-selected (r-strategists) and K-selected (K-strategists) species based on the level of parental investment in offspring and the number of offspring produced, as shown in the table below.

Strategy	r-selected	K-selected
Example	Fruit fly	Elephant
Environmental conditions	Unstable	Stable
Body size	Smaller	Larger
Generation time	Short	Longer
Breeding occasions	Few, often only once	Many more
Number of offspring	Many	Fewer
Parental care	Limited	Extensive
Offspring survival	Very low	Higher

Comparison of costs and benefits of external and internal fertilisation

In external fertilisation, the meeting of eggs and sperm occurs in watery habitats outside the body of the individual organisms, whereas internal fertilisation occurs within the body of female individuals.

A principal benefit of external fertilisation is that very large numbers of offspring can be produced at one time, but the costs are high because many gametes – which are shed into the environment – could be predated or not used in fertilisation. There is very limited or no parental care so relatively few offspring survive.

A main benefit of internal fertilisation is that because of the increased chance of successful fertilisation, fewer eggs are needed. Offspring can be retained internally for protection and/or development, resulting in a higher offspring survival rate. The costs of internal fertilisation include the need to locate a mate, which requires energy expenditure, and the difficulty of the direct transfer of gametes from one partner to the other.

Check-up 37 ?

1 Describe the greater parental investment made by females compared with males. **2**
2 Describe the costs and benefits of parental investment. **2**
3 Describe the differences between r-selection and K-selection. **4**
4 Compare internal and external fertilisation in terms of the costs and benefits of each. **6**

Reproductive behaviours and mating systems in animals

Mating systems are based on how many mates an individual has during one breeding season, as shown in Figure 2.29. These range from polygamy, in which individuals of one sex have more than one mate – this includes polygyny, in which one male mates exclusively with a group of females, and polyandry, in which one female mates with a number of males in the same breeding season – to monogamy, the mating of a pair of animals to the exclusion of all others.

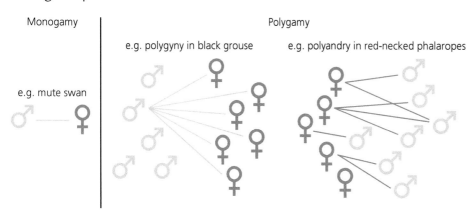

Monogamy

Polygamy

e.g. polygyny in black grouse

e.g. polyandry in red-necked phalaropes

e.g. mute swan

Figure 2.29 Examples of mating systems in some British birds

Courtship

Courtship is behaviour that leads to breeding success. Courtship behaviour in birds and fish can be a result of species-specific sign stimuli and instinctive fixed action pattern responses.

Example

Courtship in three-spined sticklebacks

The courtship behaviour of three-spined sticklebacks is an example in which the sign stimuli of a swollen belly of the female and the red underparts of the territorial male lead to a fixed action pattern of behaviour that results in the fertilisation of the eggs. Behaviours follow one from the other in a set, genetically determined sequence, each behaviour acting as a stimulus for the next, as shown in Figure 2.30.

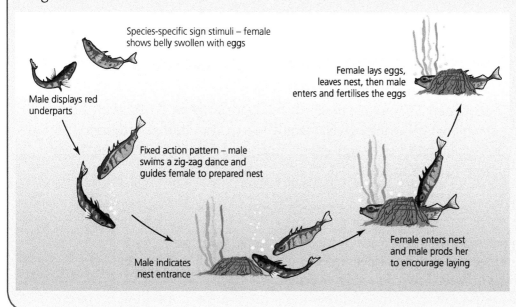

Species-specific sign stimuli – female shows belly swollen with eggs

Female lays eggs, leaves nest, then male enters and fertilises the eggs

Male displays red underparts

Fixed action pattern – male swims a zig-zag dance and guides female to prepared nest

Male indicates nest entrance

Female enters nest and male prods her to encourage laying

Figure 2.30 Courtship in sticklebacks involves species-specific sign stimuli and fixed action patterns, which are instinctive

Check-up 38

1 Describe **two** examples of mating systems.   **2**
2 Explain the difference between polygyny and polyandry.   **2**
3 Describe features common to successful courtship behaviour in birds and fish.   **2**

Sexual selection

Sexual selection selects for characteristics that have little survival benefit for the individual but increase their chances of mating. The ornate plumes and conspicuous dances of many bird-of-paradise species have evolved as a result of sexual selection. These species have evolved in habitats in which food is abundant and predators are scarce.

Many species exhibit sexual dimorphism as a product of sexual selection. Females are generally inconspicuous, often with cryptic colouration that blends in with the surroundings, while males usually have more conspicuous markings, structures and behaviours, as shown in Figure 2.11 on page 76. Reversed sexual dimorphism occurs in some species. In this system, females take on male-like roles and behaviour, and are often more conspicuous.

Key links 👍

There is more about sexual dimorphism in Key Area 2.2.

Female choice

Female choice involves females assessing honest signals of the fitness of males, often during courtship behaviour. Honest signals can indicate favourable alleles that increase fitness and, therefore, chances of survival of offspring, or a low parasite burden suggesting a healthy individual.

Some birds are lekking species, in which males gather to perform courtship displays at a specific site called a lek. For example, in black grouse in the UK (shown in Figure 2.12b) on page 76) males dance, show white tail feathers and bright red eye wattles at lekking sites. These features act as honest signals that allow females to assess males. Dominant males occupy the centre of the lek, with subordinates and juveniles at the fringes as satellite males. Female choice occurs during the display, with the dominant male gaining access to most females. The satellites benefit by the presence of females attracted by the dominant male and will attempt secretive mating with some of them, often successfully. In some species, the territorial males benefit from the presence of satellite birds since their presence seems to attract additional females to the lek.

Male–male rivalry

Success in male–male rivalry, through real or ritualised conflict, increases access to females for mating. Males will fight for dominance and access to females, often using elaborate weaponry such as antlers, tusks or horns. In red deer there is a seasonal rut in which stags fight for hinds, the winner securing the mating rights to a harem of females. Antlers play a key part in success at the rut.

Hints & tips ⭐

Remember that female choice is different from male–male rivalry, although they have similarities and both are examples of sexual selection.

Check-up 39

1 Describe what is meant by sexual dimorphism.   **1**
2 Explain what is meant by a lekking species.   **1**
3 If a display during lekking provides honest signals, explain the benefit that may be obtained by females receiving these signals.   **2**
4 Describe the characteristics of male–male rivalry in sexual selection.   **2**

Key links

There is more about female choice and male–male rivalry in Key Area 2.2.

Key words

Fixed action pattern – species-specific sequence of behaviours in which one behaviour leads to the next

Honest signals – characteristics that can indicate fitness and favourable alleles that increase the chances of survival of offspring or a low parasite burden, suggesting a healthy individual

K-selected (K-strategists) – breeding strategy in species that are larger, live longer and produce relatively few, larger offspring; they give a high level of parental care and offspring have a high probability of surviving to adulthood

Lek – a communal area in which two or more males of a lekking species perform courtship displays called lekking

Male–male rivalry – males will fight for dominance and access to females for mating through real or ritualised conflict

Monogamy – the mating of a pair of animals to the exclusion of all others

Parental investment – any parental expenditure that benefits offspring; it increases the offspring's chances of surviving and reproductive success at the expense of the parent's ability to invest in other offspring

Polyandry – one female mates with a number of males in the same breeding season

Polygamy – individuals of one sex having more than one mate

Polygyny – one male mates with a number of females in the same breeding season

Reversed sexual dimorphism – sexual dimorphism in which females are more conspicuous than males

r-selected (r-strategists) – breeding strategy in species that are smaller, mature more rapidly and produce a larger number of smaller offspring, each of which receives limited parental care and most will not reach adulthood

Sexual dimorphism – the differences in appearance between males and females of the same species, such as in colour, shape, size and structure, as a product of sexual selection

Sexual selection – selection for characteristics that have little survival benefit for the individual, but increase their chances of mating

Species-specific sign stimulus – a feature or action that indicates intent, often to start courtship

Exam-style questions ?

Structured questions

1 The ruff is a medium-sized, ground-nesting wading bird. Males display and compete for females during the breeding season at grassy sites called leks. There are two male forms: territorial males with conspicuous dark plumage, and satellite males with conspicuous light plumage. Territorial males occupy and defend the best mating territories in the lek. Satellite males don't hold territory but enter leks and attempt to mate with females visiting those territories occupied by territorial males. The presence of both types of male in a territory attracts additional females. Females are polygamous and lack the conspicuous plumage of males.

⇨

   a) State the term used to describe structural differences between males and females of the same species. **1**

   b) Explain the selective advantages that the territorial ruff and the satellite male gains in securing mating opportunities with females. **2**

   c) Suggest the selective advantage to territorial males in tolerating the presence of satellites. **1**

   d) **(i)** Describe what is meant by the term 'polygamous'. **1**

       (ii) Explain the selective advantage to female ruff of having inconspicuous plumage. **1**

2 a) Black grouse are a lekking species of the Scottish moorland. Males perform dance-like displays in which they show off their white tail feathers and bright red wattles. These are thought to be honest signals.

      (i) Explain what is meant by a lekking species. **1**

      (ii) Explain why lekking displays are thought to be examples of sexual selection. **2**

      (iii) If the display contains honest signals, describe the information females may get from these and explain the advantage of the signals to them in mate selection. **2**

   b) Red deer are also a moorland species. They have an annual rut in which male stags fight for mating rights with female hinds.

      (i) Explain how a rut differs from a lek. **2**

      (ii) Suggest two factors that may make a stag successful in a rut. **2**

Extended response

3 Compare and contrast r-selected and K-selected species. **8**

4 Describe courtship behaviours that affect reproductive success. **8**

Answers are given on pages 156–157.

Key Area 2.5
Parasitism

Key points !

1   An ecological **niche** is a multi-dimensional summary of tolerances and requirements of a species. ☐

2   A species has a **fundamental niche** that it occupies in the absence of any interspecific competition. ☐

3   A **realised niche** is occupied in response to interspecific competition.

4   As a result of interspecific competition, **competitive exclusion** can occur, where the niches of two species are so similar that one declines to local extinction. ☐

5   Where the realised niches are sufficiently different, potential competitors can co-exist by **resource partitioning** in which resources are split between potential competitors. ☐

6   Symbiosis is an intimate co-evolved relationship between two species that can be notated + for benefit, – for detriment and 0 for neutral. ☐

7   Parasitism is a symbiotic interaction between a parasite and its host in which the parasite benefits (+) in terms of nutrients at the expense of its host, which loses these (–). ☐

8   Unlike a predator–prey relationship, the reproductive potential of a parasite is greater than that of its host. ☐

9   Most parasites have a narrow, specialised niche as they are very host-specific. ☐

10 As the host provides so many of the parasite's needs, many parasites are degenerate, lacking structures and organs such as digestive systems found in other organisms. ☐

11 An **ectoparasite** lives on the surface of its host, whereas an **endoparasite** lives within the tissues of its host. ☐

12 Some parasites require only one host to complete their life cycle, while others require more than one host to complete their life cycle.

13 The **definitive host** is the organism on or in which the parasite reaches sexual maturity. ☐

14 **Intermediate hosts** may also be required for the parasite to complete its life cycle. ☐

15 A **vector** plays an active role in the transmission of the parasite and may also be a host. ☐

16 The human disease malaria is caused by the parasite *Plasmodium*.

17 An infected mosquito, acting as a vector, bites a human and *Plasmodium* enters the bloodstream; asexual reproduction occurs in the liver and then in the red blood cells, which burst to release **gametocytes** into the bloodstream. ☐

18 If a mosquito bites an infected human, gametocytes enter the mosquito, maturing into male and female gametes and allowing sexual reproduction to occur. ☐

19 Schistosome parasites cause the human disease schistosomiasis. ☐

20 Schistosomes reproduce sexually in the human intestine and fertilised eggs pass out via faeces into water where they develop into larvae that infect water snails, where asexual reproduction occurs producing another type of motile larvae that escape the snail and penetrate the skin of humans, entering the blood stream. ☐

21 Viruses are parasites that can only replicate inside a host cell. ☐

22 Viruses contain genetic material in the form of DNA or RNA, packaged in a protective protein coat. ☐

23 Some viruses are surrounded by a phospholipid membrane derived from host cell materials. ☐

24 The outer surface of a virus contains antigens that a host cell may or may not be able to detect as foreign. ☐

25 Viral life cycle stages include infection of the host cell with genetic material, then host cell enzymes replicate the viral genome, which is transcribed and translated into viral proteins; new viral particles are assembled and released from host cells. ☐

26 **RNA retroviruses** use the enzyme reverse transcriptase to form DNA, which is then inserted into the genome of the host cell, allowing new viral particles to be formed. ☐

27 **Transmission** is the spread of a parasite to a host. ☐

28 **Virulence** is the harm caused to a host species by a parasite. ☐

⇨

29 Ectoparasites are generally transmitted through direct contact or by consumption of intermediate hosts. ☐

30 Endoparasites of the body tissues are often transmitted by vectors.

31 Factors that increase transmission rates include the overcrowding of hosts when they are at high density, and mechanisms such as vectors and waterborne dispersal stages, which allow the parasite to spread even if infected hosts are incapacitated. ☐

32 Host behaviour is often exploited and modified by parasites to maximise transmission. ☐

33 Alteration of host foraging, movement, sexual behaviour, habitat choice or anti-predator behaviour becomes part of the **extended phenotype** of the parasite. ☐

34 Parasites often suppress the host immune system and modify host size and reproductive rate in ways that benefit parasite growth, reproduction or transmission. ☐

35 Immune response in mammals involves both **non-specific defences** and **specific cellular defences**. ☐

36 Physical barriers, chemical secretions, **inflammatory response**, **phagocytes** and **natural killer cells** destroying cells infected with viruses are examples of non-specific defences. ☐

37 Skin and epithelial tissue acts as physical barriers by blocking the entry of parasites. ☐

38 Chemical secretions include **hydrolytic enzymes** in mucus, saliva and tears, which destroy bacterial cell walls; secretions of the stomach, vagina and sweat glands create low pH environments that denature cellular proteins of pathogens. ☐

39 Injured cells release signalling molecules that enhance blood flow to the site, bringing antimicrobial proteins and phagocytes. ☐

40 Phagocytes can kill parasites using powerful enzymes contained in lysosomes, by engulfing them and storing them inside a vacuole in the process of phagocytosis. ☐

41 Natural killer cells can identify and attach to cells infected with viruses, releasing chemicals that lead to cell death by inducing apoptosis. ☐

42 A range of white blood cells constantly circulate in the blood, monitoring the tissues. ☐

43 If tissues become damaged or invaded, cells release cytokines that increase blood flow, resulting in non-specific and specific white blood cells accumulating at the site of infection or tissue damage. ☐

44 Mammals contain many different lymphocytes, each possessing a receptor on its surface that can potentially recognise a parasite antigen. ☐

45 Binding of an antigen to a lymphocyte's receptor selects that lymphocyte to divide and produce a clonal population of this lymphocyte. ☐

46 Some **selected lymphocytes** produce antibodies while others can induce apoptosis in parasite-infected cells. ☐

47 Antibodies possess regions where the amino acid sequence varies greatly between different antibodies; this variable region gives the antibody its specificity for binding antigen. ☐

48 When the antigen binds to this binding site the antigen–antibody complex formed can result in inactivation of the parasite, rendering it susceptible to a phagocyte, or it can stimulate a response that results in cell lysis. ☐

49 Initial antigen exposure produces **memory lymphocyte** cells specific for that antigen that can produce a secondary response when the same antigen enters the body in the future. ☐

50 In a secondary response, antibody production is enhanced in terms of speed of production, concentration in the blood and duration. ☐

51 Parasites have evolved ways of evading the immune system. ☐

52 Endoparasites mimic host antigens to evade detection and modify the host immune response to reduce their chances of destruction. ☐

53 **Antigenic variation** in some parasites allows them to change between different antigens during the course of infection of a host, and may also allow reinfection of the same host with the new variant. ☐

54 Some viruses escape immune surveillance by integrating their genome into host genomes, existing in an inactive state known as **latency**. ☐

55 A virus emerges from latency when favourable conditions arise.

56 **Epidemiology** is the study of the outbreak and spread of infectious disease. ☐

⇨

57 The **herd immunity** threshold is the density of resistant hosts in the population required to prevent an epidemic. ☐

58 **Vaccines** contain antigens that will elicit an immune response. ☐

59 The similarities between host and parasite metabolism makes it difficult to find drug compounds that only target the parasite. ☐

60 Antigenic variation has to be reflected in the design of vaccines.

61 Some parasites are difficult to culture in the laboratory, making it difficult to design vaccines. ☐

62 Challenges arise where parasites spread most rapidly as a result of overcrowding or tropical climates. ☐

63 Overcrowding can occur in refugee camps that result from war or natural disasters, or rapidly growing cities in less economically developed countries (**LEDCs**). ☐

64 Overcrowded conditions make co-ordinated treatment and control programmes difficult to achieve. ☐

65 Civil engineering projects to improve sanitation, combined with co-ordinated vector control, may often be the only practical parasite control strategies. ☐

66 Improvements in parasite control reduce child mortality and result in population-wide improvements in child development and intelligence, as individuals have more resources for growth and development. ☐

Concept of niche

An ecological niche is a multi-dimensional summary of tolerances and requirements of a species. A species has a fundamental niche that it occupies in the absence of any interspecific competition. A realised niche is occupied in response to interspecific competition, as shown in Figure 2.31a). As a result of interspecific competition, competitive exclusion can occur, where the niches of two species are so similar that one declines to local extinction. Where the realised niches are sufficiently different, potential competitors can co-exist by resource partitioning, as shown in Figure 2.31b).

a)

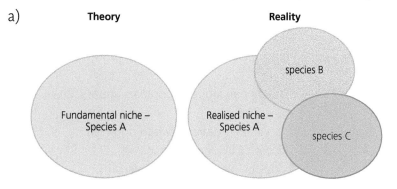

b)

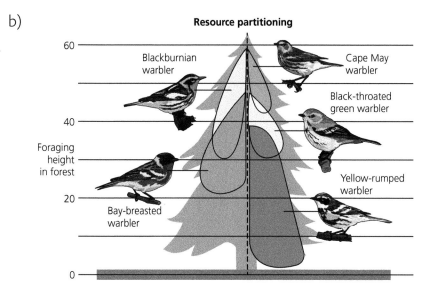

Figure 2.31 a) Fundamental and realised niche; b) an example of resource partitioning in New World warblers – although all five species live together in woodland, they partition resources by feeding in different but overlapping stations

The parasite niche

Symbiosis is a co-evolved and intimate relationship between species. Parasitism is a symbiotic interaction between a parasite and its host, which can be described using a +/−/0 notation. The parasite benefits (+) in terms of nutrients at the expense (−) of its host. Unlike a predator–prey relationship, the reproductive potential of the parasite is greater than that of the host. Most parasites have a narrow, specialised niche as they are very host-specific. As the host provides so many of the parasite's needs, many parasites are degenerate, lacking the structures and organs found in other organisms.

An ectoparasite lives on the surface of its host, whereas an endoparasite lives within the tissues of its host.

> **Key link** 👍
>
> There is more about the +/− notation in Key Area 2.2.

> **Check-up 40**
>
> 1 Describe the difference between the fundamental and realised niche of a species.   **2**
> 2 Explain the principles of competitive exclusion and resource partitioning.   **2**
> 3 Describe the characteristics of the parasitic niche.   **4**

Parasitic life cycles

Some parasites require only one host, but many parasites require more than one host to complete their life cycle. The definitive host is the organism on or in which the parasite reaches sexual maturity. Intermediate hosts may also be required for the parasite to complete its life cycle. A vector plays an active role in the transmission of the parasite and may also be a host.

The malaria parasite

The human disease malaria is caused by an endoparasite called *Plasmodium*. An infected mosquito, acting as a vector and the definitive host, bites a human – the intermediate host – as shown in Figure 2.32. *Plasmodium* enters the human bloodstream in saliva in the bite. Asexual reproduction of the parasite occurs in the human liver and then in the red blood cells. The red blood cells burst and gametocytes are released into the bloodstream. Another mosquito bites the infected human and the gametocytes enter its body, maturing into male and female gametes allowing fertilisation and so sexual reproduction to occur. The zygote develops in the mosquito's gut and stages pass into its salivary glands so that the mosquito can then infect another human host when it bites.

The schistosomiasis parasite

Schistosomes cause the human disease schistosomiasis. Schistosomes reproduce sexually in the human intestine, making humans the primary host. Their fertilised eggs pass out via faeces into water where they develop into larvae. The larvae then infect water snails, the intermediate host, where asexual reproduction occurs producing another type of motile larvae. These then escape the snail into the water and penetrate the skin of humans wading in the water and entering their bloodstream, as shown in Figure 2.33.

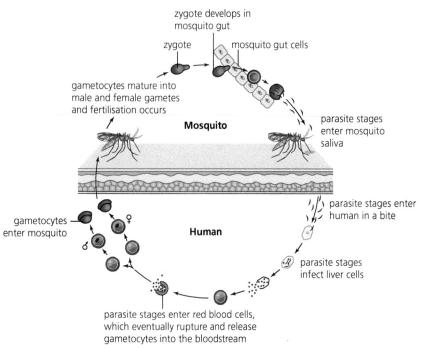

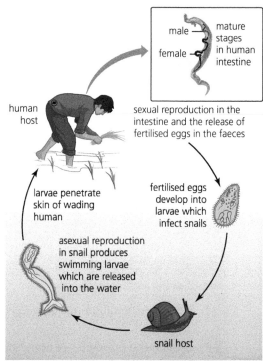

Figure 2.32 Life cycle of *Plasmodium* – note that the mosquito is a vector as well as the definitive host because it actively transmits the parasite

Figure 2.33 Life cycle of schistosomes – note that the snail is an intermediate host but not a vector because it does not actively transmit the parasite

Viruses

Viruses are parasites that can only replicate inside a host cell. They contain their genetic material in the form of DNA or RNA, packaged in a protective protein coat. Some viruses are surrounded by a phospholipid membrane derived from host cell materials. The outer surface of a virus contains antigens that host cells may or may not be able to detect as foreign. Viruses attach to the host cell and it can become infected with their genetic material. The host cell and viral enzymes replicate the viral genome. Viral genes can be transcribed and translated to produce viral proteins including coats. Replicated viral genes and newly formed coats can assemble to form new viral particles that can be released by the host cell, as shown in Figure 2.34. These viruses can go on to infect new host cells.

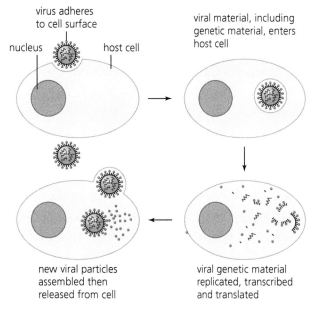

Figure 2.34 Stages in the infection cycle of a typical virus

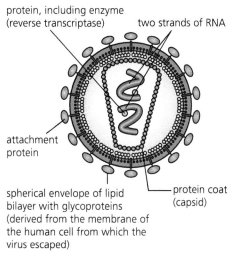

Figure 2.35 Details of a single RNA retrovirus – this shows a section through an HIV particle

RNA retroviruses, as shown in Figure 2.35, have genes made of strands of RNA. The enzyme reverse transcriptase transcribes it into DNA, which is then inserted into the genome of the host cell.

Check-up 41

1 Describe a definitive host and explain why many parasites require more than one host.   **2**
2 Explain the role of a vector in a parasitic life cycle.   **1**
3 Describe the stages in the life cycle of *Plasmodium*.   **5**
4 Describe the structure of a virus.   **2**
5 Describe the stages of a viral life cycle.   **5**
6 Describe the replication of RNA retroviruses.   **3**

Transmission and virulence

Transmission is the spread of a parasite to a host. Virulence is the harm caused to a host species by a parasite. Ectoparasites are generally transmitted through direct contact or by consumption of intermediate hosts. Endoparasites of the body tissues are often transmitted by vectors. Factors that increase transmission rates of parasites include overcrowding of hosts when they are at high density, and the presence of abundant vectors. Waterborne dispersal stages allow the parasite to spread even if infected hosts are incapacitated.

Host behaviour is often exploited and modified by parasites to maximise transmission. The host behaviour becomes part of the extended phenotype of the parasite, which includes modification of host foraging, movement, sexual behaviour, habitat choice or anti-predator behaviour, to maximise transmission. Parasites often suppress the host immune system and modify host size and reproductive rate in ways that benefit the parasite's growth, reproduction or transmission.

Check-up 42

1 Describe the transmission of ectoparasites and endoparasites.   **2**
2 Describe **two** factors that increase parasite transmission rates.   **2**
3 Describe a modification of host behaviour that can be seen as an extended phenotype and how it can be an advantage to the parasite.   **3**

Defence against parasitic attack

The immune response to parasitic attack in mammals has both non-specific and specific aspects. Non-specific defences are shown in the following table.

Non-specific defence	Examples
Physical barriers	Epithelial tissue blocks the entry of pathogens
Chemical secretions	Hydrolytic enzymes in mucus, saliva and tears destroy bacterial cell walls; low pH environments created by the secretions of the stomach, vagina and sweat glands denature the cellular proteins of pathogens
Inflammatory response	Injured cells release signalling molecules that result in enhanced blood flow to the site, bringing antimicrobial proteins and phagocytes
Phagocytes	Engulf pathogens into a vacuole and release powerful enzymes contained in lysosomes, as shown in Figure 2.36
Natural killer cells	Can identify and attach to cells infected with viruses, releasing chemicals that lead to cell death by inducing apoptosis

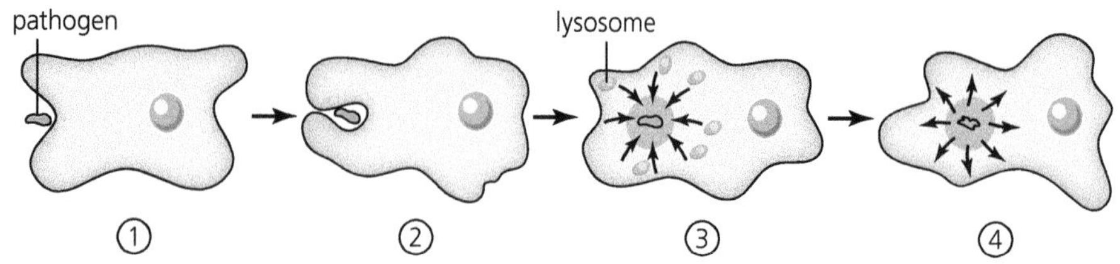

pathogen

lysosome

① ② ③ ④

Key links 👍

There is more about lysosomes in Key Area 1.2.

Figure 2.36 Phagocytosis:①phagocyte detects pathogen;②pathogen engulfed into a vacuole;③ lysosomes fuse with vacuole and add digestive enzymes;④pathogen destroyed and any waste materials diffuse into cytoplasm

Check-up 43 ❓

1 Describe **two** examples of the action of chemical secretions in the defence against parasitic attack.   **4**
2 Describe the inflammatory response.   **3**
3 Describe the action of phagocytes in defence against parasitic attack.   **3**
4 Describe the role of natural killer cells in the defence against parasitic attack. **3**

Specific defences

In mammals, a range of white blood cells constantly circulate, monitoring the tissues. If tissues become damaged or invaded, cells release cytokines that increase blood flow, resulting in non-specific and specific white blood cells accumulating at the site of infection or tissue damage. Mammal tissues contain many different lymphocytes, each possessing a receptor on its surface that has the potential to recognise a specific parasite antigen. Binding of an antigen to a lymphocyte's receptor selects that lymphocyte to then divide and produce a clonal population of this lymphocyte, as shown in Figure 2.37.

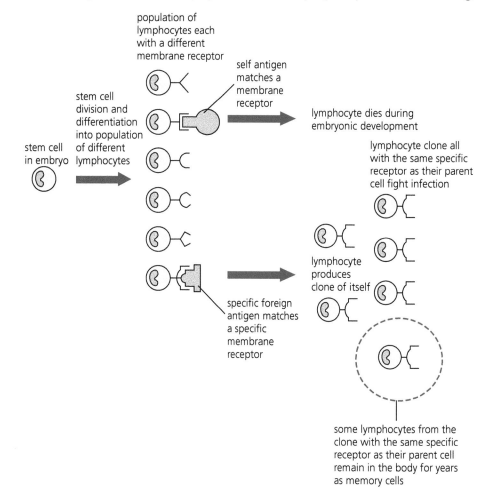

Figure 2.37 Clonal selection of lymphocytes

Some selected B lymphocytes will produce antibodies while others can induce apoptosis in parasite-infected cells. Antibodies possess regions where their amino acid sequence varies greatly between different antibodies, as shown in Figure 2.38. It is this variable region that gives the antibody its specificity for antigen binding. When the antigen binds to this binding site, the antigen–antibody complex formed can result in inactivation of the parasite, rendering it susceptible to a phagocyte, or it can stimulate a response that results in cell lysis.

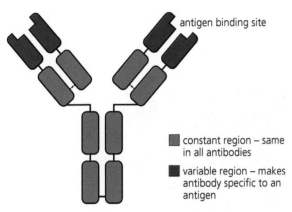

Key links

There is more about apoptosis in Key Area 1.5.

Figure 2.38 Structure of a specific antibody showing the variable region where the specific antigen binding site is developed

Memory lymphocyte cells are also formed as shown in Figure 2.37. Initial antigen exposure produces memory lymphocyte cells specific for that antigen and that can produce a secondary response when the same antigen enters the body in the future. When this occurs, antibody production is enhanced in terms of speed of production, concentration in blood and duration, as shown in Figure 2.39.

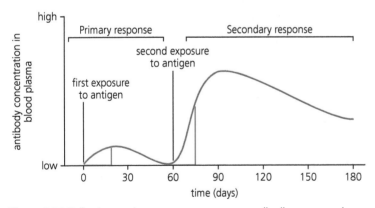

Figure 2.39 Following a primary response, memory cells allow a secondary response that is faster, higher and of longer duration than the primary response

Check-up 44

1 Describe the clonal selection of lymphocytes in a specific response to parasites.   **2**
2 Explain the specificity of antibodies.   **2**
3 Describe **two** possible processes by which lymphocyte clones defend the body.   **2**
4 Describe the features of the secondary immune response.   **2**

Immune evasion

Parasites have evolved ways of evading the immune system. Endoparasites mimic host antigens to evade detection and modify the host's immune response to reduce their chances of destruction.

Antigenic variation in some parasites allows them to change between different antigens during the course of infection of a host. It may also allow reinfection of the same host with the new variant.

Some viruses escape immune surveillance by integrating their genome into host genomes, existing inside cells in an inactive state known as latency. The virus becomes active again when favourable conditions arise.

Challenges in treatment and control

Epidemiology is the study of the outbreak and spread of infectious diseases. The herd immunity threshold is the density of resistant hosts in the population required to prevent an epidemic, as shown in Figure 2.40. It can be challenging to achieve herd immunity in less economically developed countries (LEDCs) and in developed countries in which the population resists vaccination. Vaccines contain antigens that will elicit an immune response.

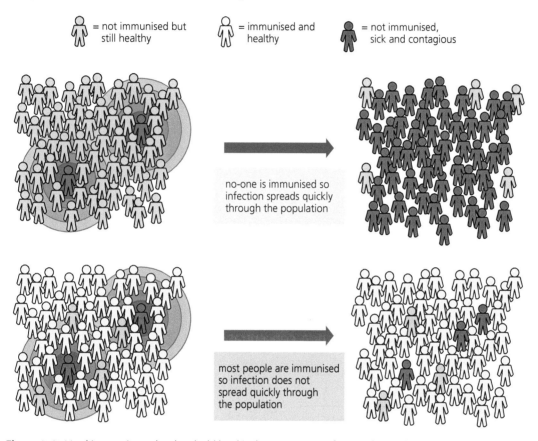

Figure 2.40 Herd immunity – the threshold level is the percentage of a population that has to be immunised to prevent an epidemic

> **Hints & tips** ⭐
>
> *Make sure you know that epidemiology is the study of outbreaks of diseases and their spread.*

Problems with drugs, vaccines and social conditions

The similarities between host and parasite metabolism makes it difficult to find drug compounds that target the parasite only. Vaccines contain specific antigenic material to cause an immune response but do not themselves cause disease. Antigenic variation has to be reflected in the design of these vaccines, and some parasites are difficult to culture in the laboratory, making it difficult to design vaccines.

Challenges arise where parasites spread most rapidly as a result of overcrowding or tropical climates. Overcrowding can occur in refugee camps that result from war or natural disasters, or rapidly growing cities in LEDCs. These conditions make co-ordinated treatment and control programmes difficult to achieve. Civil engineering projects to improve sanitation combined with co-ordinated vector control may often be the only practical control strategies.

Improvements in parasite control reduce child mortality and result in population-wide improvements in child development and intelligence, as individuals have more resources for growth and development.

Check-up 45 ?

1	Describe how endoparasites have evolved to evade the immune system.	**3**
2	Describe what is meant by herd immunity.	**1**
3	Explain why it is difficult to find drug compounds that only target the parasite.	**1**
4	Describe what is meant by a vaccine and explain why they can be difficult to design.	**2**
5	Describe the challenges in the treatment and control of parasites.	**4**

Key words

Antigenic variation – change between different antigens during the course of infection of a host

B lymphocytes – white blood cells that produce specific antibodies in response to specific antigens

Competitive exclusion – where the niches of two species are so similar that one declines to local extinction

Definitive host – the organism on or in which the parasite reaches sexual maturity

Ectoparasite – parasite that lives on the surface of its host

Endoparasite – parasite that lives within the tissues of its host

Epidemiology – the study of the outbreak and spread of infectious diseases

Extended phenotype – the expression of a parasite's genotype into the phenotype of its host by manipulating the host phenotype to facilitate its transmission

Fundamental niche – the niche that is occupied in the absence of any interspecific competition

Gametocytes – the precursors of male and female gametes

Herd immunity threshold – the density of resistant hosts in the population required to prevent an epidemic

Hydrolytic enzymes – any enzyme that catalyses the hydrolysis of a chemical bond

Inflammatory response – injured or wounded areas become warm and red due to increased blood flow, bringing white cells for defence

Intermediate host – a host that is normally used by a parasite in the course of its life cycle and in which it may multiply asexually but not sexually

Latency – viruses escape immune surveillance by integrating their genome into host genomes, existing in an inactive state

LEDC – less economically developed country

Memory lymphocyte – lymphocyte specific for a specific antigen; they are retained in the body following infection and can produce a secondary response to the same antigen

Natural killer cells – lymphocytes responsible for destroying abnormal cells

Niche – a multi-dimensional summary of tolerances and requirements of a species

Non-specific defences – general response to infection, including phagocytosis

Phagocyte – white blood cell in non-specific defence, engulfing and destroying foreign antigens; may also present antigens to lymphocytes

Realised niche – the niche that is occupied in response to interspecific competition

Resource partitioning – where the realised niches are sufficiently different that potential competitors can co-exist

RNA retroviruses – viruses that use the enzyme reverse transcriptase to form DNA

Specific cellular defences – activity of the immune system in response to a particular pathogen, triggered by antigens located on the surface of cells

Transmission – the spread of a parasite to a host

Vaccines – contain antigens that will elicit an immune response

Vector – an organism that does not cause disease itself but which spreads the parasite from one host to another

Virulence – the harm caused to a host species by a parasite

Exam-style questions

Structured questions

1   The diagram below shows the distribution of two species of barnacle on a rocky shore.

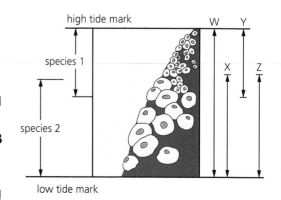

The line Z shows the realised niche of species 2.

a) Define the term 'niche'. **1**

b) In terms of species 1 and 2, identify the niches shown by lines W, X and Y. **3**

c) The two species can co-exist by the partitioning of resources.
Explain what is meant by this statement. **1**

2   The diagram below shows the life cycle of a parasitic schistosome that causes schistosomiasis in humans.

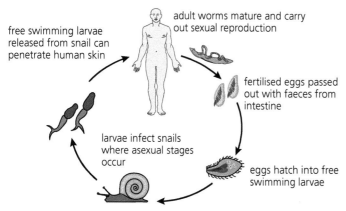

a) Explain why:
   (i)   the human can be described as the definitive host **1**
   (ii) the snail cannot be described as a vector. **1**

b) (i)   Describe a feature of the life cycle of the flatworm that can increase its rate of transmission. **1**
   (ii) Suggest **two** methods of controlling this parasite. **2**

c) Give **two** ways in which parasites can evade detection by host immune systems. **2**

Extended response

3   Give an account of defence against parasitic attack under the following headings:
   a) non-specific defences **4**
   b) specific defences. **6**

4   Give an account of endoparasitic infections under the following headings:
   a) problems with treatments and the prevention of transmission **4**
   b) other parasite control methods and advantages to human population of parasite control. **3**

Answers are given on page 158.

Practice course assessment: organisms and evolution

Section 1

1 The arctic fox is a predator of barnacle geese. To reduce predation, geese are periodically vigilant, which means they look up from time to time while grazing to scan for foxes. In a study of this behaviour, different flock sizes of geese were monitored for the ten-minute period after a model fox was placed 100 m from the feeding flock. The percentage of time each individual spent with their head raised was recorded and the results are show in the graph below.

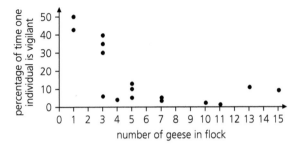

To test the hypothesis that vigilance behaviour of individual barnacle geese in response to the presence of foxes decreases as flock size increases, the work should be repeated

A and average results calculated

B and recording time increased to 20 minutes

C but data for missing flock sizes should be obtained

D but have trials with no model fox present.

2 Which line in the table correctly shows the meanings of terms used in the study of animal behaviour?

| | Terms used in the study of animal behaviour | | |
	Latency	Frequency	Duration
A	Time between the presentation of a stimulus and the behaviour it evokes	Number of times a behaviour is performed in a set period	Time over which a behaviour occurs
B	Time over which a behaviour occurs	Number of times a behaviour is performed in a set period	Time between the presentation of a stimulus and the behaviour it evokes
C	Time between the presentation of a stimulus and the behaviour it evokes	Time over which a behaviour occurs	Number of times a behaviour is performed in a set period
D	Time over which a behaviour occurs	Number of times a behaviour is performed in a set period	Time between the presentation of a stimulus and the behaviour it evokes

3 Which of the following situations would be expected to increase the rate of evolution?

A having a longer generation time

B living in a cooler environment

C reducing selection pressure

D transferring genes horizontally

4 In terms of selection, fitness can be described as absolute or relative. Absolute fitness is the ratio of:

A surviving offspring of one phenotype compared to other phenotypes

B frequencies of a particular genotype after selection compared with before

C surviving offspring of one genotype compared to the most successful genotype

D frequencies of a particular phenotype in one generation compared to the next.

5 The diagram opposite shows some details of the body of an individual nematode worm *Caenorhabditis elegans*.

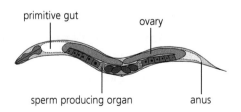

The information given suggests that this individual is:

A parasitic   C parthenogenic

B hermaphroditic   D sexually dimorphic.

6 Which line in the table below correctly describes cells at the end of a meiotic stage?

	Meiotic stage	Chromosome complement of cells present at end of stage	Number of cells at end of stage
A	Meiosis I	Haploid	4
B	Meiosis I	Diploid	2
C	Meiosis II	Haploid	4
D	Meiosis II	Diploid	2

7 The red-necked phalarope, *Phalaropus lobatus*, is a ground-nesting wading bird. The females have brighter plumage than the males and the males carry out much of the egg incubation.
This situation is described as:

A satellite male strategy   C lekking behaviour

B reversed sexual dimorphism   D parthenogenesis.

8 During ritualised courtship in peafowl, *Pavo muticus*, the male spreads and shakes his tail feathers to attract a female before stepping back and bowing. This is followed by loud mating calls.
This type of fixed action pattern response can be a result of:

A honest signals   C male–male rivalry

B sexual dimorphism   D species-specific sign stimuli.

9 The fundamental niche of a species:

A includes the set of resources available in the absence of competition

B includes the set of resources available in the presence of competition

C permits co-existence in a community

D permits the sharing of resources with other species.

10 The diagram opposite shows a response by B lymphocytes to foreign antigens.
Which of the following correctly identifies this cellular response?

A apoptosis   C inflammation

B clonal selection   D phagocytosis

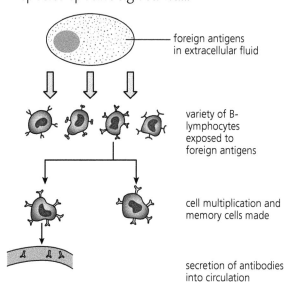

foreign antigens in extracellular fluid

variety of B-lymphocytes exposed to foreign antigens

cell multiplication and memory cells made

secretion of antibodies into circulation

Section 2

1 The Lincoln Index is used to estimate the size (N) of certain animal populations during field investigations. A sample of the population is captured and marked (M). After an appropriate time, a second sample is captured (C) and any recaptured individuals are counted (R).

a) Give the formula that is used for estimating the population size.   1

b) Give **two** assumptions that need to be made when using this method of mark and recapture for a valid and reliable estimate to be made.   2

c) Give **one** method of marking the animals being monitored.   1

d) Give **one** reason why scientists monitor certain populations of organisms.   1

e) Name an important ethical requirement that must be considered when sampling wild populations.   1

2 Female parasitic wasps, *Nasonia vitripennis*, lay their eggs inside the pupae of houseflies, *Musca domestica*. The wasp eggs hatch into larvae that consume the housefly pupae.

In a study to investigate the evolutionary response of the host to the parasite, two containers were set up with housefly populations. Container A had a housefly population with no previous exposure to the parasite and Container B had a housefly population which had been exposed to wasp parasitism for a period of three years prior to the study.

The graphs below show how the populations of each species in the containers changed over a 40-week period.

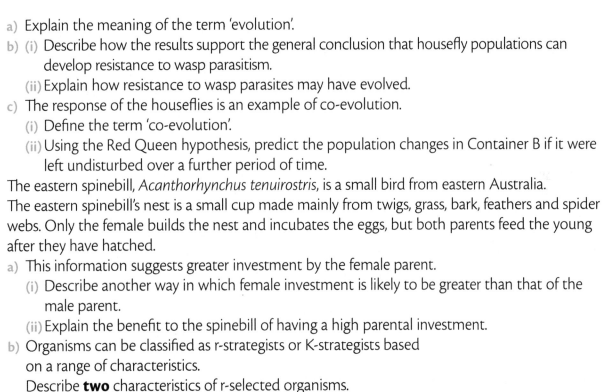

a) Explain the meaning of the term 'evolution'. **1**

b) (i) Describe how the results support the general conclusion that housefly populations can develop resistance to wasp parasitism. **1**

  (ii) Explain how resistance to wasp parasites may have evolved. **2**

c) The response of the houseflies is an example of co-evolution.

  (i) Define the term 'co-evolution'. **1**

  (ii) Using the Red Queen hypothesis, predict the population changes in Container B if it were left undisturbed over a further period of time. **1**

3 The eastern spinebill, *Acanthorhynchus tenuirostris*, is a small bird from eastern Australia. The eastern spinebill's nest is a small cup made mainly from twigs, grass, bark, feathers and spider webs. Only the female builds the nest and incubates the eggs, but both parents feed the young after they have hatched.

a) This information suggests greater investment by the female parent.

  (i) Describe another way in which female investment is likely to be greater than that of the male parent. **1**

  (ii) Explain the benefit to the spinebill of having a high parental investment. **1**

b) Organisms can be classified as r-strategists or K-strategists based on a range of characteristics.

Describe **two** characteristics of r-selected organisms. **2**

c) The red-winged blackbird, *Agelaius phoenicius*, is widely distributed in large areas of North and Central America. Territorial males show a form of polygamy and may mate with and defend up to ten females in the territory, depending on the food resources available.

  (i) Name the form of polygamy shown by the male red-winged blackbird. **1**

  (ii) Explain why a male bird should favour this form of mating system. **1**

4 The common dandelion, *Taraxacum officinale*, is a plant species that shows geographic parthenogenesis. The plants can be parasitised by rust fungi of the order *Pucciniales*.

Populations of dandelions along a line of increasing latitude in central Europe, as shown in the map below, were studied. The percentages of parthenogenic plants and the incidence of infection by rust fungi were determined and the results are shown in the table.

111

Transect point	% of parthenogenic plants	Incidence of parasitic rust infection (% plants infected)
1	18	48.0
2	15	54.0
3	48	3.0
4	100	0.0
5	100	5.0

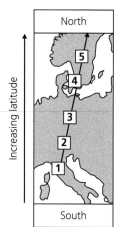

a) State what is meant by the term 'parthenogenesis'.   **1**

b) Explain the relationship between the parasite, parthenogenesis and the latitude of populations of dandelion.   **2**

c) Vegetative cloning in plants is an example of asexual reproduction in eukaryotes.

    (i) Give **one** advantage and **one** disadvantage of asexual reproduction as a reproductive strategy.   **2**

    (ii) State **one** factor that can increase variation in asexually reproducing populations.   **1**

5 Schistosoma is a parasitic flatworm found in tropical areas throughout the world. The flatworm can live for many years within a host. In humans, if untreated, it causes the disease schistosomiasis (bilharzia), which can be fatal. Adult parasites pass fertilised eggs into the large intestine, which pass out via faeces into water. The parasite's eggs hatch in fresh water, releasing a free-living stage that infects a species of freshwater snail. The parasite multiplies asexually within this host before being released into the water as a second free-living stage. This stage is capable of penetrating the skin of humans and other mammals when they are in fresh water. Inside the liver of the mammal, the flatworms develop into sexually mature adults that disperse eggs via the host's large intestine. Successful control of Schistosoma is very difficult.

a) Describe the features of parasitism as a symbiotic interaction.   **1**

b) State which of the two hosts involved in this life cycle may be described as the definitive host. Justify your answer.   **1**

c) Explain why the life cycle stage that involves the freshwater snails benefits the parasite.   **1**

d) Suggest **one** reason why successful control of Schistosoma has proved very difficult to achieve.   **1**

e) Parasites living inside a host will be exposed to attack by the host's immune system. Describe **one** way in which parasites may overcome the immune response of their hosts.   **1**

f) Many endoparasites lack structures and organs found in other organisms. Give the term for organisms adapted in this way.   **1**

6 a) Describe the events of meiosis and how they lead to variation in gametes.   **10**

    OR

b) Describe the basis of sex determination in animals.   **10**

Answers can be found on pages 158–159.

Area 3 Investigative biology

Scientific principles and process

Key points !

1 The scientific cycle involves observation; the construction of a testable **hypothesis**; experimental design; the gathering, recording and analysis of data; the evaluation of results and conclusions; and the formation of a revised hypothesis where necessary. ☐

2 In science, refinement of ideas is the norm, and scientific knowledge can be thought of as the current best explanation, which may then be updated after evaluation of further experimental evidence. ☐

3 The **null hypothesis** (H_0) proposes that there will be no statistically significant effect as a result of the experiment treatment. ☐

4 Failure to find an effect (a negative result) is a valid finding, as long as an experiment is well designed. Conflicting data or conclusions can be resolved through careful evaluation or can lead to further experimentation. ☐

5 If there is evidence for an effect, unlikely due to chance, then the null hypothesis is rejected. ☐

6 Scientific ideas only become accepted once they have been checked independently. ☐

7 Effects must be reproducible; one-off results are treated with caution. ☐

8 Publication of methods, data, analysis and conclusions in scientific reports is important so that others are able to repeat an experiment. ☐

9 Common methods of sharing original scientific findings include seminars, talks and posters at conferences, and publishing in academic journals. ☐

10 **Peer review** of scientific publications and critical evaluation by specialists with expertise in the relevant field are essential to the scientific process. ☐

11 Most scientific publications use peer review. Specialists with expertise in the relevant field assess the scientific quality of a submitted manuscript and make recommendations regarding its suitability for publication. ☐

12 **Review articles**, which summarise current knowledge and recent findings in a particular field, are essential to scientists undertaking new research in that field. ☐

13 Science coverage in the wider media requires critical evaluation before it should be accepted. ☐

14 There has been increasing public understanding of science, and the issue of misrepresentation of science has been raised. ☐

15 The unbiased presentation of results, citing and providing references, and avoiding **plagiarism** are important. ☐

16 While judgements and interpretations of scientific evidence may be disputed, integrity and honesty are of key importance in science. ☐

17 The **replication** of experiments by others reduces the opportunity for dishonesty or the deliberate misuse of science. ☐

18 In animal studies, the concepts of **replacement**, **reduction** and **refinement** are used to avoid, reduce or minimise harm to animals. ☐

19 Informed consent, the right to withdraw and confidentiality are key factors in human studies. ☐
⇨

20 The value or quality of science investigations must be justifiable in terms of the benefits of its outcome, including the pursuit of scientific knowledge. ☐

21 As a result of the risks involved, many areas of scientific research are highly regulated and licensed by governments. ☐

22 The risk to and safety of subject species, individuals, investigators and the environment must be taken into account. ☐

23 Legislation, regulations, policy and funding can all influence scientific research. ☐

24 Legislation limits the potential for the misuse of studies and data. ☐

The scientific cycle

Science works in a methodical way following a tried and tested method called the scientific cycle, as shown in Figure 3.1. Observation of the biological world leads to questions that require explanation. Biological science uses agreed steps to approach these questions and reach answers.

You should be aware that laboratory experimentation and field investigation both follow similar cycles. Laboratory experiments are performed largely *in vitro*, where control of variables is possible but the application of results to life can be tricky. Field investigations are carried out *in vivo*, where control of variables is tricky and maybe impossible, but the application of results to life is clear.

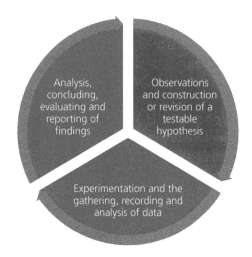

Figure 3.1 The scientific cycle – there is no end to the cycle and it continually turns to produce increasingly better explanations of the observed biological world; most of the ideas covered in this chapter fall into the red segment of the cycle

Key links

There is more about *in vitro* and *in vivo* science in Key Area 3.2.

Constructing hypotheses

A hypothesis is a prediction of how a dependent variable would change when an independent variable is altered experimentally. A hypothesis should be unambiguous and able to be tested and measured.

The null hypothesis (H_0) proposes that there will be no statistically significant effect on the dependent variable as a result of the experimental manipulation of independent variables. It is a valuable hypothesis for the scientific method because it is easy to test using statistical analysis, which means that the hypothesis can be supported or rejected with a high level of confidence. Failure to find an effect is a negative result and is a valid finding, as long as the experiment is well designed. If there is evidence for an effect that is thought unlikely to have occurred by chance, then the null hypothesis can be rejected. Rejection of H_0 is almost expected in some investigations and can lead to an alternative hypothesis (H_1) being constructed and tested.

At any given time in history there is a currently accepted set of ideas that explain our biological observations of nature. The on-going refinement of ideas is accepted as normal, and scientific knowledge should be thought of only as the current best explanations. Ideas may be updated after evaluation of further experimental evidence.

Check-up 46 ❓

1	Describe the order of the main events in the scientific cycle.	**4**
2	Explain the null hypothesis (H_0) and its usefulness.	**2**

Conflict and acceptance

In many cases experimentation produces data that is conflicting or conclusions that are not accepted by all. These issues can be resolved through careful evaluation of experimental methods and data, leading to further experimentation as appropriate. Scientific ideas should only become accepted once they have been checked independently, so effects must be reproducible and one-off results treated with caution.

The importance of publication

Scientists generally publish their work as papers in academic journals that adhere to strict common standards when assessing papers for publication. Papers are expected to follow a standard structure:

- introduction – puts the work into context, and includes its aims and any hypothesis
- method – describes the procedures followed and the materials and instrumentation used
- results – section in which data is presented and analysed
- discussion – in which conclusions are drawn and evaluations of various aspects of the work included.

Sharing original scientific findings

Common methods of sharing original scientific findings include publishing in academic journals as well as the giving of seminars and talks at conferences and symposia. Science coverage in the wider media – such as television broadcasts, the popular press and magazines – should also be critically evaluated by producers and editors, as well as viewers and readers.

Review articles summarise current knowledge and recent findings in a particular field. They are usually written by leading scientists in that particular area.

Ethical considerations

Ethics deals with issues of right and wrong, and it impacts scientists personally as well as their treatment of the organisms that they work with.

Personal ethics

Reputable journals ensure that all published papers have been peer reviewed and critically evaluated. Integrity and honesty of scientists, an unbiased presentation of results, citing and providing references, and avoiding plagiarism, are crucial and are also subject to editorial scrutiny and peer review. While judgements and interpretations of scientific evidence may be disputed, integrity and honesty are of key importance in science. The replication of experiments by others reduces the opportunity for dishonesty or the deliberate misuse of science.

Animal studies

In animal studies, the concepts of replacement, reduction and refinement are used to avoid, reduce or minimise the harm to animals:

- Replacement involves using alternatives such as cell or tissue samples instead of whole organisms.
- Reduction involves changing experimental design or statistical methods to lower the number of individual animals required.
- Refinement involves the consideration of how experiments are carried out to ensure the minimum pain or harm, and also includes improvements in animal welfare such as accommodation, food supply and veterinary provision.

Informed consent and the right to withdraw data, as well as confidentiality, are key features of scientific work involving human studies.

Check-up 47

1 Give **two** ways in which science findings are published. **2**
2 Name the basic headings under which a scientific paper would be written. **5**
3 Describe what is meant by a review article. **2**
4 Give **three** methods by which harm can be minimised when working with natural populations. **3**

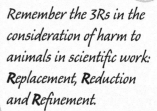

Hints & tips

Remember the 3Rs in the consideration of harm to animals in scientific work: Replacement, Reduction and Refinement.

Key words

Hypothesis– prediction of a relationship between independent and dependent variables

Null hypothesis– easy to test prediction that there is no relationship between the independent and dependent variable

Peer review– appraisal of a piece of work or a scientific report by an independent expert in the relevant field

Plagiarism– copying the work of others without acknowledgement

Reduction– lowering the numbers of an organism used in a study in order to minimise harm to a species

Refinement– the consideration and adoption of methods to minimise harm to a species, including improved animal accommodation and veterinary provision

Replacement– use of an alternative to whole organisms in a study to minimise harm to a species

Replication– repeating an experiment either within the study or independently to improve reliability

Review article– expert report that summarises all that is known about an area of interest

Exam-style questions

Structured question

1 Arrange the following parts of the scientific cycle into the correct order: **5**

analysis of data	hypothesis formation	gathering of data	experimental design
experimentation	observation of nature	evaluating and concluding	publication

Extended response

2 Give an account of the main methods used to reduce harm to animals involved in scientific research. **6**

Answers are given on page 160.

Key Area 3.2a

Experimentation – experimental skills

Key points !

1 **Validity** relates to **variables** that are controlled so that any measured effect is likely to be due to the **independent variable**. ☐

2 Variables can be **continuous** or **discrete**. ☐

3 **Reliability** is related to confidence in the data and obtaining consistent values in repeats and independent replicates. ☐

4 **Accuracy** relates to how close to the true value data, or means of data sets, are. ☐

5 **Precision** relates to how close measured values are to each other. ☐

6 A **pilot study** is used to help plan procedures, assess validity and check techniques; it is integral to the development of an investigation. ☐

7 A pilot study allows the evaluation and modification of an experimental design. ☐

8 The use of a pilot study can ensure an appropriate range of values for the independent variable and can help establish the number of repeat measurements required to give a representative value for each independent datum point. ☐

9 An independent variable is the variable that is changed in a scientific experiment. ☐

10 A **dependent variable** is the variable being measured as the results of a scientific experiment. ☐

11 Experiments involve the manipulation of an independent variable by the investigator. ☐

12 The experimental treatment group is compared to a control group. ☐

13 A simple experimental design is one in which there is one independent variable; a **multifactorial** experimental design has more than one independent variable or a combination of treatments. ☐

14 Investigators may use experimental groups that already exist, so there is no truly independent variable. ☐

15 The control of laboratory conditions allows simple experiments to be conducted more easily than in the field; a drawback is that findings may not be applicable to a wider setting. ☐

16 **Observational studies** are good at detecting **correlation**, but since they do not directly test a hypothesis, they are less useful for determining **causation**. ☐

17 In observational studies the independent variable is not directly controlled by the investigator, for ethical or logistical reasons. ☐

18 Due to the complexities of biological systems, **confounding variables** (other variables besides the independent variable) may affect the dependent variable. ☐

19 Confounding variables must be held constant if possible, or at least monitored so that their effect on the results can be accounted for in the analysis. ☐

20 In cases where confounding variables cannot be controlled easily, a **randomised block design** could be used. ☐

21 Randomised blocks of treatment and control groups can be distributed in such a way that the influence of any confounding variable is likely to be the same across treatment and control groups. ☐

22 Control results are used for comparison with the results of treatment groups – **negative** and **positive controls** may be used. ☐

23 A negative control provides results in the absence of a treatment. A positive control is a treatment that is included to check that the system can detect a positive result when it occurs. ☐

⇨

⇨

24 **Placebos** lacking the independent variable being investigated can be included as a treatment in human trials. ☐

25 The placebo effect is a measurable change in the dependent variable as a result of a human patient's expectations, rather than changes in the independent variable. ☐

26 *In vitro* refers to the technique of performing a given procedure in a controlled environment outside of a living organism. ☐

27 Examples of *in vitro* experiments include cells growing in culture medium, proteins in solution and purified organelles. ☐

28 Advantages of *in vitro* studies are that they are controllable, repeatable, rapid, cheap and may avoid ethical and regulatory issues. ☐

29 Disadvantages of *in vitro* studies are that extrapolation to life is difficult and chronic effects are not tested. ☐

30 *In vivo* refers to experimentation using a whole, living organism. ☐

31 Advantages of *in vivo* studies are that they simulate real life and chronic effects can be tested. ☐

32 Disadvantages of *in vivo* studies are that they are difficult to control, are slow and expensive, and may have ethical and regulatory issues. ☐

33 Where it is impractical to measure every individual in a population, a **representative sample** of a population is selected. ☐

34 The extent of the natural variation within a population determines the appropriate sample size. ☐

35 More variable populations require larger sample sizes. ☐

36 A representative sample should share the same mean and the same degree of variation about the mean as the population as a whole. ☐

37 In **random sampling**, members of the population have an equal chance of being selected. ☐

38 In **systematic sampling**, members of a population are selected at regular intervals because of an environmental gradient. ☐

39 In **stratified sampling**, a non-homogeneous population is divided into categories called strata that are then sampled proportionally. ☐

40 Variation in experimental results may be due to the reliability of measurement methods and/or inherent variation in the specimens. ☐

41 The reliability of measuring instruments or procedures can be determined by repeated measurements or readings of an individual datum point; the variation observed indicates the precision of the measurement instrument or procedure, but not necessarily its accuracy. ☐

42 The natural variation in the biological material being used can be determined by measuring a sample of individuals from the population. ☐

43 The mean of repeated measurements gives an indication of the true value being measured. ☐

44 The **range** of values is a measure of the extent of variation in the results. ☐

45 If there is a narrow range then the variation is low. ☐

46 Independent replication involves repeating experimental procedures with a different investigator, or in a different laboratory, or at a different time; it is carried out to produce **independent data sets**. ☐

47 Overall results can only be considered reliable if they can be achieved consistently. ☐

48 Independent data sets should be compared to determine the reliability of the results. ☐

Experimental design in the scientific cycle

Experimental design is related to the carrying out of experiments and the gathering and recording of data, as shown in the red-coloured segment of Figure 3.2.

Experiments are usually conducted under laboratory conditions, but field investigations are conducted under the prevailing field conditions. Account must be taken of the crucial principles that give confidence in science: validity, reliability and accuracy and precision of data. These principles are explained in the table below.

Figure 3.2 The experimental design area of the scientific cycle is shown in red

Validity	Reliability	Accuracy	Precision
Variables controlled so that any measured effect is likely to be due to the independent variable – this is easier to achieve in the laboratory than in the field	Consistent data obtained in repeats and in replicates carried out independently	Accurate data are close to the true values for the measurement made	Precise data values in replicated measurements lie close to each other

low accuracy
low precision

low accuracy
high precision

high accuracy
low precision

high accuracy
high precision

Figure 3.3 The difference between accuracy and precision illustrated; accuracy refers to the closeness of a measurement to a standard value (the bullseye of the target in this example), while precision refers to the closeness of two or more measurements to each other

 *Hints & tips*

In your exam there will be a large experimental question worth between 5 and 9 marks, but there will be other questions on experimental design throughout the paper.

Variables

An independent variable is the variable that is changed or manipulated in a scientific experiment. The investigator changes this variable then measures the effects it has. The dependent variable is the variable that is measured to show the effects that an independent variable might have on it.

Experimental variables may be continuous or discrete. Continuous variables are measured with instruments and can show an infinite range of values from the minimum to the maximum value of the variable. An example of a continuous variable is the temperature of the atmosphere measured using a thermometer. Discrete variables can only be expressed in whole numbers. An example of a discrete variable is the number of a particular species of animal counted in a point count.

Controls

So that an investigator can be confident that it is the independent variable that is having an effect on the dependent variable, all other possible variables must be controlled by trying to keep them constant. This is an essential step in making an experimental procedure valid. Laboratory conditions are easier to control than a field investigation.

In experiments involving an independent variable in the form of a treatment, such as a fertiliser or drug for example, a negative control group that does not receive the treatment is often used to further highlight the role of the independent variable in changes to the dependent variable. Placebos are negative control treatments without the presence of the independent variable being investigated and are often used in human drug testing trials (for example, a tablet, pill or injection that does not contain the drug being trialled).The placebo effect is a measurable change in the dependent variable as a result of a human patient's expectations, rather than changes in the independent variable.

Positive controls can be set up to check that the experimental system is able to detect a positive result when it occurs. Control results are used for comparison with the results of treatment groups.

In vitro refers to the technique of performing a given procedure in a controlled laboratory environment outside of a living organism. Examples of *in vitro* experiments include cells growing in culture medium, studying proteins in solution and conducting experiments on purified organelles. The advantages of *in vitro* studies are that they are controllable, repeatable, rapid, cheap and may avoid ethical and regulatory issues. The disadvantages are that extrapolation to whole organisms or natural ecosystems is difficult and chronic effects are not tested.

In vivo refers to experimentation using whole living organisms in the laboratory or in nature. Advantages of *in vivo* studies are that they simulate real life and chronic effects can be tested. The disadvantages are that they are difficult to control, are slow and expensive and may have ethical and regulatory issues.

Experimental design

A simple experimental design is one in which there is only one independent variable. In a multifactorial experimental design, there is more than one independent variable or a combination of treatments. For example, if new fertilisers are being tested, it is possible that they will contain different concentrations of different minerals – each mineral acting as an independent variable. The results of multifactorial

---

Hints & tips

When explaining the purpose of a treatment with the independent variable at zero, it is worth using the term 'negative control' as part of your answer.

Hints & tips

Be aware that placebo effects are limited to humans and are unlikely to be seen in animal subjects.

Check-up 48

1 Explain what is meant by validity and reliability in science.   **4**
2 Describe the difference between accuracy and precision in measurements.   **2**
3 Describe the difference between continuous and discrete variables.   **2**
4 Describe the difference between *in vitro* and *in vivo* investigations in terms of the advantages and disadvantages they have.   **4**
5 Explain the differences between negative and positive control.   **2**
6 Describe the placebo effect.   **2**

experiments are more difficult to analyse and make valid conclusions from. Investigators may decide to use groups that already exist. For example, if they were investigating the effects of peppered moth wing colouration on survival; the wing colours already exist so there is no true independent variable. This type of independent variable is said to be attributed or passive but can be treated as if it were truly independent.

Randomised block design

In cases where confounding variables cannot easily be controlled, a randomised block design could be used. Randomised blocks of treatment and control groups can be distributed in such a way that the influence of a confounding variable is likely to be the same across the treatment and control groups. In the field trial shown in Figure 3.4, six blocks have been created randomly on a hillside with a west to east prevailing wind in each block. A crop is grown in four different plots: three with different fertiliser treatments and a control plot. This design also builds in some replication.

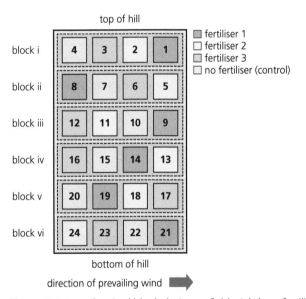

Figure 3.4 A randomised block design to field trial three fertilisers and minimise the effects of any potentially confounding variables that are introduced by the slope of the land or the direction of the wind

Pilot studies and sampling

A pilot study is used to help plan procedures, assess validity and check techniques, and is integral to the development of an investigation. It allows evaluation and modification of experimental design, and can be used to check that the most appropriate range of values for an independent variable is used, or to decide the level that controlled variables should be held at. It also helps to establish the number of repeat measurements required to give a representative value for each independent datum point.

Check-up 49 ❓

1 Give **two** examples of independent variables which already exist in wild populations so are not truly independent. **2**

2 Describe how simple experiments differ from multifactorial ones. **2**

3 Explain the advantages of randomised block designs in field trials. **2**

In most cases, it is likely to be impractical to measure every individual in a population, and so a representative sample of the population is selected. The extent of the natural variation within a population helps to determine the appropriate sample size. More variable populations require larger sample sizes. A representative sample should share the same mean and the same degree of variation about the mean as the population as a whole. Samples can only be representative if they are free from selection bias, which can arise through inappropriate sampling methods or those that may not be truly random. Sampling methods should be appropriate to the situation and the hypotheses being tested.

The examples in Figure 3.5 and the table below represent different field situations and appropriate sampling methods. In each case twelve samples are collected, but the methodology of selection is different.

a)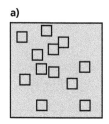

b)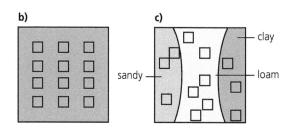

c)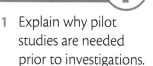

Figure 3.5 Different sampling strategies to achieve representative sampling depending on plant species and its observed distribution in the area: a) random; b) systematic along transects; c) stratified by soil type

Diagram	Sampling method	Habitat	Description
a)	Random	Homogeneous, e.g. a ploughed cornfield	Randomly selected sampling stations; all members of the population have an equal chance of being selected
b)	Systematic	Environmental gradient, e.g. a steep hillside	Sampling stations at regular intervals along a transect
c)	Stratified	Heterogeneous, e.g. presence of different soil types	Population divided into groups or strata based on knowledge of habitat then sampled proportionately

- In random sampling, all members of the population have an equal chance of being selected in the sample. This method might be used for an apparently homogeneous habitat, such as a ploughed cornfield.
- In systematic sampling, members of a population are selected at regular intervals, such as at the stations along a transect line crossing a habitat where there is an environmental gradient, for example a hillside.
- In stratified sampling, a non-homogeneous population is divided into categories called strata that are then sampled proportionally. For example, the habitat can be divided into different soil types before undertaking sampling.

Check-up 50

1 Explain why pilot studies are needed prior to investigations.   2

2 Explain what is meant by a representative sample.   2

3 Describe the differences between systematic and stratified sampling techniques.   2

Key links 👍

There is more about sampling and transects in Key Area 2.1.

Reliability

Variation in experimental results may be due to the reliability of measurement methods and/or inherent variation in the individuals sampled. The reliability of measuring instruments or procedures can be determined by repeated measurements or readings of an individual datum point. The variation observed in the results of this indicates the precision of the measurement instrument or procedure, but not necessarily its accuracy.

Hints & tips

Make sure you know the difference between precision and accuracy.

The natural variation of the biological material being used can be determined by measuring a sample of individuals from the population. The mean of these repeated measurements will give an indication of the true value being measured. The range of values is a measure of the extent of variation in the results. If there is a narrow range, then the variation is low.

Independent replication should be carried out to produce independent data sets. Independent replication is not simply repetition. Replication is best achieved when a procedure or experiment is done by another person, preferably in a different laboratory at a different time. Repetition occurs when multiple sets of measurements are made during one scientific investigation.

Overall results can only be considered reliable if they can be achieved consistently, and these independent data sets should be compared to determine the reliability of the results.

Check-up 51 ?

1   Describe the factors that can give rise to variation in experimental data.   **2**
2   Explain the difference between a repeated measurement and an independently replicated measurement.   **2**

Key words

Accuracy – the degree of closeness to the true, actual value of measurement

Causation – a link between variables in which one variable is known to be directly affecting the other

Confounding variables – factors that influence the results of an experiment and cause mistaken associations between the independent and dependent variable to be made

Continuous variable – variable that can be measured and for which infinite values exist

Correlation – a relationship between two variables that does not imply causation

Dependent variable – factor that is measured to obtained experimental data

Discrete variable – variable that must fall into clear-cut categories

In vitro – experimental procedure carried out in laboratory conditions using parts of organisms such as cells or tissues

In vivo – procedures carried out in laboratory or field conditions using entire, living organisms

Independent data sets – repeated data sets that are collected in different labs by different workers and at different times but under the same general conditions

Independent variable – experimental variable that is purposely altered by the investigator

Multifactorial – an experimental situation in which there is more than one dependent variable

Negative control – an experimental aspect in which the independent variable is set at zero, or at no treatment, with the aim of producing a negative result

Observational studies – work that is usually carried out in the field or *in vivo*, and usually produces qualitative data

Pilot study – a small-scale study conducted to refine values for independent and controlled variables prior to conducting an experiment

Placebos – negative controls used in drug and vaccine trials that do not contain the active ingredient being tested

Positive control – an experimental aspect set up to show that a system is capable of detecting a positive result should it occur

Precision – the closeness of repeated measurements of a variable

Random sampling – sampling in such a way as to ensure that all individuals have an equal chance of being selected in order to obtain a statistically representative sample

Randomised block design – experimental protocol in which the effects of potentially confounding variables can be reduced

Range – the difference between the two extremes of a set of numerical data

Reliability – the degree of confidence that an experimental procedure can produce consistent values

Representative sample – a sample that shares the same mean and same degree of variation about the mean as the population as a whole

Stratified sampling – dividing a population into groups or strata before carrying out the sampling to take account of perceived differences in the individuals, such as size or age

Systematic sampling – sampling at regular intervals in space or time to take account of a gradient, such as a slope, a tidal cycle or seasonal changes

Validity – refers to the control of variables to produce fair testing

Variable – factor in an experiment that is changeable or can change

Exam-style questions

Structured questions

1   Phosphorylase is an enzyme found in potato tuber cell cytoplasm. It converts glucose-1-phosphate (G1P) to starch, which is stored in grains in the tuber cells.

In an experiment aiming to compare concentration of the enzyme, starch-free phosphorylase extracts were prepared from random samples of the tubers of three different potato varieties using identical preparation procedures. Standard volumes of the extracts were mixed with standard volumes of G1P, and the time taken for starch molecules to be formed was measured. Several replicates for each potato variety were done and the results averaged. As a positive control, a solution of commercially available phosphorylase was also tested.

The results are shown in the table below.

Phosphorylase source	Average time to convert 1 cm³ of 1% G1P to starch
Variety 1 extract	240
Variety 2 extract	230
Variety 3 extract	270
Commercial phosphorylase solution	150

    a) (i)   Identify the independent variable in this experiment.   **1**

       (ii) Identify a potentially confounding variable in this experiment.   **1**

    b) (i)   Explain what is meant by a positive control.   **1**

       (ii) Suggest how a negative control for this experiment could be designed.   **1**

    c)   Explain why it was important to ensure that the extracts were free of starch.   **1**

d) (i) Potato varieties were sampled randomly.
Suggest how stratified sampling of the potato varieties could be done and how that
would improve experimental validity. **2**
(ii) Comment on the reliability of the experimental method. **1**

2 An investigation was carried out into the effects of different concentrations of ATP on lean muscle
tissue taken from a sample from a young lamb. Thin strips of muscle were cut from the sample and each
carefully laid out on glass microscope slides. The length of each strip was measured and recorded. Equal
volumes of ATP solution of different concentrations were added to the strips, as shown in the table
below. After five minutes the lengths of the muscle strips were recorded as shown.

Concentration of ATP solution (%)	Strip	Initial length of strip (mm)	Final length of strip (mm)	Change in length (mm)
0 (control)	1	10.5	10.5	0
	2	10.0	9.5	0.5
	3	9.5	9.5	0
5	1	11.0	10.5	0.5
	2	10.5	9.5	1.0
	3	9.5	9.0	0.5
10	1	10.0	8.5	1.5
	2	10.5	8.0	2.5
	3	11.0	9.0	2.0
15	1	10.5	7.0	3.5
	2	11.5	8.5	3.0
	3	11.0	7.0	4.0

a) Identify the independent variable in this experiment. **1**
b) Two potentially confounding variables in this experiment are the temperatures of the solutions
and muscle strips during the experiment, and the breed of sheep that the samples came from.
(i) Suggest **one** further possible confounding variable in this experiment. **1**
(ii) Explain **one** way in which this variable could affect the result of the experiment. **2**
c) State whether or not the data shown is reliable and explain your answer. **2**
d) Identify the type of control used in this experiment. **1**
e) Suggest how selection bias has affected the validity of this experiment. **1**

Extended response

3 Give an account of the principles and strategies that should be used to obtain representative samples. **4**
4 Give an account of experimental design under the following headings:
a) multifactorial experiments **3**
b) randomised blocks. **4**

Answers are given on page 161.

Key Area 3.2b
Experimentation – data handling skills

Key points !

1 Variables can give rise to quantitative, qualitative or ranked data. ☐
2 **Quantitative data** are obtained objectively with instruments and have numerical values. ☐
3 **Qualitative data** are subjective, descriptive and based on observation. ☐
4 **Ranked data** refers to the data transformation in which numerical values are replaced by their rank when the data are sorted from lowest to highest. ☐
5 The type of variable being investigated determines the graphical display or statistical tests that may be used. ☐
6 The arithmetic **mean** is the average of a data group; the **median** is the middle value, with equal numbers of data below and above it; the **mode** is the most common value or category. ☐
7 **Box plots** are used to show variation within and between data sets. ☐
8 Box plots of data sets show the median, the lower quartile, the upper quartile and the inter-quartile range. ☐
9 **Error bars** on graphical data indicate the range of the data. ☐
10 **Correlation** exists if there is a relationship between two variables visible in data. Correlation is an association and does not imply **causation**. ☐
11 Causation exists if the changes in the values of the independent variable are known to cause changes to the value of the dependent variable. ☐
12 A positive correlation exists when an increase in one variable is accompanied by an increase in the other variable. ☐
13 A negative correlation exists when an increase in one variable is accompanied by a decrease in the other variable. ☐
14 Correlations can be strong or weak. ☐
15 Strength of correlation is proportional to the spread of values from a **line of best fit**. ☐

Data handling in the scientific cycle

Hints & tips ★

In your exam there will be a large data handling question worth between 7 and 10 marks, but there will be other questions on data handling through the paper.

Data handling is related to selecting and analysing data from experiments and field investigations; it is represented by the red-coloured segment of the scientific cycle shown in Figure 3.6.

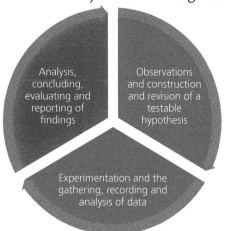

Analysis, concluding, evaluating and reporting of findings

Observations and construction and revision of a testable hypothesis

Experimentation and the gathering, recording and analysis of data

Figure 3.6 The data handling area of the scientific cycle is shown in red

Types of data

Data is information related to an investigation. Quantitative data has numerical values. The pH of river water or the temperature of soil are quantitative. Qualitative data is descriptive and based on observation. The gender of a fruit fly or the colour of a peppered moth's wings are qualitative.

Ranked data is produced when data values are transformed into ranked categories. The order of attendance of different species of vulture at a carcass, or the allocation of bird numbers to categories such as abundant, common or rare, are examples of ranked data.

The type of variable and data collected determines the graphical presentation of the data, as shown in the table and Figure 3.7 below. The statistical tests that are used subsequently also depend on the types of variables and data.

Data	Description	Example	Graph type often used
Qualitative	Fairly common type of data that is subjectively observed or described	Animal behaviour categories, colour of peppered moth wings	Bar chart, pie chart
Quantitative	Very common type of data that is objectively measured usually with instrumentation	Temperature, pH, time, concentration, body mass	Scatter plot, line graph, histogram, box plot
Ranked	Unusual type of data with values placed into hierarchical order	Results of population estimates involving allocation of numerical data to abundance ranks	Bar chart, stacked bar chart

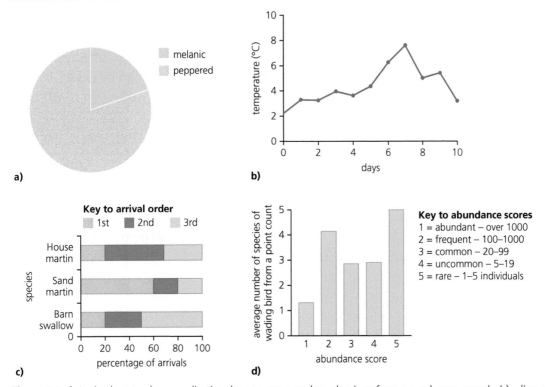

Figure 3.7 a) A pie chart to show qualitative data on peppered moth wings from a moth trap sample; b) a line graph to show quantitative data of the temperature of a river over a ten-day period; c) a stacked bar chart to show ranked data of the order of arrival of three closely related species of migrant bird to their breeding habitat in Scotland over a ten-year period; d) a bar chart showing the species of wading birds from a point count ranked by their abundance

Measures of central tendency in the data set

Data sets can be processed to give statistical values intended to give representations of the complete data set. Some of these are measures of central tendency. The arithmetic mean is the average of all the values in the data set. The median is the numerically middle value in data, so that half the data values are below it and half above it. The mode is the most common value in the data set.

Biological variation generally shows a normal distribution of values, as shown in Figure 3.8a). In that example the mean, median and mode have the same value. Other biological data may not follow that distribution, however, and could be skewed as shown in Figure 3.8b); in that case the mean, median and mode would have different values as shown.

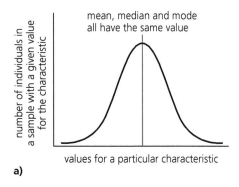

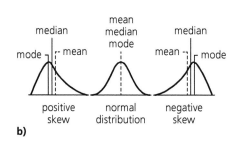

a)

b)

Figure 3.8 a) A normal distribution; b) different skews that can be seen in some data

Mean is a very useful indication of an average value in a data set but it can give the wrong impression because it can be affected badly by a small number of extreme values; the median is not affected by these so much. Also, the mean is only good for numerical data, whereas the mode can be used for qualitative data, which is non-numerical.

Measures of variation in the data set

Standard deviation

Figure 3.9 shows how data can have the same mean, median and mode but can have different spreads of variation. The standard deviation is a statistic that indicates the spread and can be used to give added descriptive quality to the measures of central tendency.

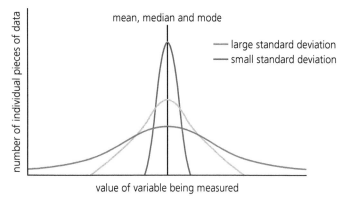

Figure 3.9 Comparison of data with the same mean, median and mode but different standard deviations

Standard error of the mean

Standard error of the mean is a statistic that is the standard deviation of a set of sample means from within a whole data set. If 100 observations were made over a set of 10 samples, each containing 10 observations, an overall mean could be calculated; then, each set of 10 observations could have its own mean calculated. The standard deviation of these means from the overall mean is the standard error of the mean.

Error bars

Error bars are lines drawn through points plotted in a graph. They may simply describe the variation of the data by indicating its range or standard deviation. Alternatively, they may indicate where the true mean of the data would be expected to lie as shown by the use of standard error of the mean or confidence interval error bars. Normally, the calculation would give a 95% chance that the true mean would lie somewhere along the error bar drawn.

Check-up 53

1 Describe the differences between mean, median and mode values. **3**
2 Describe examples of data sets for which the mean value might not be the most appropriate to reflect the data as a whole. **2**
3 Explain what the size of the standard deviation shows about the pattern of variation in data. **2**

Figure 3.10 shows the use of error bars in line graphs and bar charts. Graphs should indicate how the error bars have been calculated but, in practice, they often don't. The degree of overlap of two error bars can be an indication of the significance of the difference between their data points.

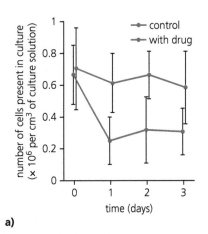

a)

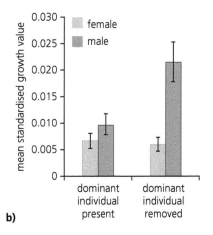

b)

Figure 3.10 a) Standard error of the mean error bars used in a line graph that compares cell growth in a culture with and without a drug under test; b) error bars used in a bar chart comparing growth values of male and female clownfish over a two-week period when a dominant individual was removed from a group and a control in which the dominant individual was not removed

Hints & tips

You would not be asked to calculate and draw your own error bars in the exam, but you do need to know about them and why they are used.

Check-up 54

1 Look at Figure 3.10a). Use the error bars to comment on the significance of the differences between the experimental drug effects and the control. **3**
2 Look at Figure 3.10b). Use the error bars to comment on:
  a) the significance of the difference in growth of males and females when a dominant individual is present **1**
  b) the significance of the removal of a dominant individual on the growth of both males and females. **2**

If error bars of two points are compared with each other as shown in Figure 3.11, their degree of overlap may tend to indicate how significantly different the two points are.

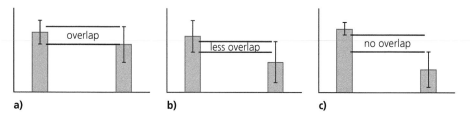

a)   b)   c)

Figure 3.11 a) When standard deviation error bars overlap a lot, it's a clue that the difference is not statistically significant; (b) when standard deviation error bars overlap a little, it's a clue that the difference is probably not statistically significant; c) when standard deviation error bars do not overlap, it's a clue that the difference may be statistically significant, but in each case a statistical test would be needed to draw a reliable conclusion

Comparing data sets using box plots

When sets of data are to be compared, a box plot is often used. The data is quantitative but is ordered to provide an overview of the variability, as shown in Figure 3.12. In this example, 15 willow warblers were trapped in June and another 15 in October at various sites in Scotland. The birds were weighed in grams to give quantitative data, but then the masses of birds in each sample were ranked according to their magnitude, as shown in Figure 3.12a). The ordered data were split into four sets, each with 25% of the data values. The highest 25% is the upper quartile and the lowest is the lower quartile, while the middle 50% is the interquartile range and contains the median value. The data can be displayed as box plots, as shown in Figure 3.12b), to show the variation within and between the data sets.

a)

Bird	Masses of individual willow warblers in sample (g)	
	June sample	Autumn sample
1	7	7
2	7	9
3	7	10
4	8	11
5	8	11
6	8	11
7	8	11
8	9	11
9	9	12
10	10	13
11	10	14
12	11	14
13	11	15
14	12	15
15	12	16
Mean	9	12

b)

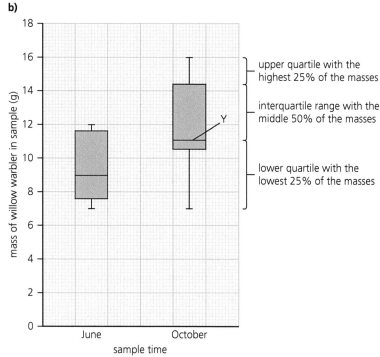

Figure 3.12 a) Raw quantitative data for the masses of two samples of willow warblers arranged in order of the magnitude of their mass; b) the same data displayed as box plots; note that Y is the median value in the data for October

Correlation

A correlation indicates a relationship between variables. Correlations and their directions can be seen in scatter plots, as shown in Figure 3.13. The lines of best fit shown are calculated from the coordinates of the data points. The strength of the correlation depends on how closely the data points are clustered along the line of best fit.

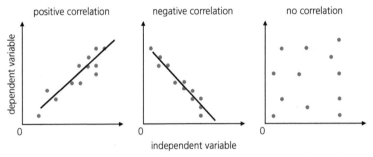

Figure 3.13 Scatter plots to indicate correlations between variable with line of best fit

> **Hints & tips**
>
> *Make sure that you realise that the upper and lower quartile lines shown on a box plot are not error bars (see Figure 3.12 above), although they look a bit like them.*

Correlation does not necessarily indicate causation because there could be a confounding variable causing the pattern of data. The correlation can only be proved if the independent variable is definitely known to be causing the changes in the dependent variable, and this can only be true if all possible variables have been controlled. While this is theoretically possible in a laboratory experiment, it is difficult or impossible in field studies.

Key words

Box plot – graph showing a data set arranged into numerical order and divided into an upper quartile, an interquartile range and a lower quartile

Causation – refers to a variable known to be causing changes in a dependent variable

Confounding variable – a variable over which no control has been possible and which may be causing changes in a dependent variable

Correlation – a link between variables in which changes in one variable bring about changes in the other

Error bar – line through a data point drawn parallel to an axis showing the variation in the data for that point or the extent of the data for which there is a 95% expectation that the true mean lies along it

Line of best fit – straight line drawn through a scatter plot that indicates the trend shown by the data

Mean – measure of central tendency obtained by summing data and dividing by the number of individual items of data

Median – measure of central tendency obtained by identifying the middle value of a data set

Mode – measure of central tendency obtained by identifying the most common value in a data set

Qualitative data – data with descriptive values

Quantitative data – data with numerical values

Ranked data – data that has been transformed into arbitrary groups

Standard deviation – value given for the spread or variation in data

Standard error of the mean – value for the standard deviation of sample means to the overall mean of a data set

Exam-style questions

Structured questions

1 The red grouse, *Lagopus lagopus scoticus*, is a game bird that occupies moorland habitats in Scotland. The birds may be infected by the endoparasitic helminth worm *Trichostrongylus tenuis*. Adult worms live in the birds' guts, and larval stages of the worm pass out in faeces. Mature larvae climb heather stems and may be eaten by grouse, which then become infected. The size of the grouse population in a study area was recorded each spring over a ten-year period between 1976 and 1986. Samples of the birds were caught in the autumn of each year and the average number of worms per bird determined. The results are shown in **Figure 1**.

Half of the birds captured in each autumn sample were treated with a drug to remove the parasitic worms, and the other half were left untreated to act as a control. All birds in the samples were weighed and marked before being released. Some of these birds were recaptured the following spring and reweighed before being released again. The results are shown in **Table 1**.

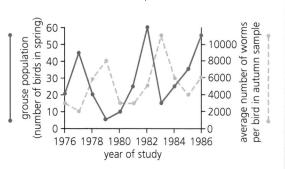

Figure 1

Sample of grouse	Average mass (kg)
Birds caught in autumn	0.720
Treated birds recaptured the following spring	0.815
Untreated birds recaptured the following spring	0.756

Table 1 Average mass of red grouse in samples

The breeding success of birds marked in 1981 was determined over the following three years by counting the average number of chicks they produced, as shown in **Figure 2**.

A computer simulation model of the effects on grouse and parasite populations of treating different percentages of grouse populations in spring with the drug was undertaken. **Figure 3** shows the results of this simulation.

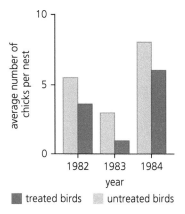

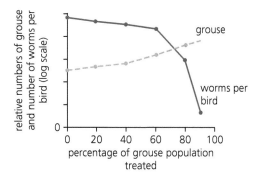

Figure 3

Figure 2

    a) It has been hypothesised that the population cycles of red grouse in Scotland are caused by the effect of the parasitic worm infection.

       (i) Describe evidence from **Figure 1** that supports this hypothesis.   **1**

       (ii) Describe additional types of evidence that would be needed to provide further support to the hypothesis.   **1**

    b) Describe the effect of the drug treatment on the mass of grouse shown in **Table 1** and suggest an explanation for this effect.   **2**

    c) (i) Describe the effect of infection with parasitic worms on the reproductive success of the grouse shown in **Figure 2** and suggest an explanation for this effect.   ⇨ **2**

⇨

(ii) Give **one** piece of evidence that indicates that worm infection is not the only factor that affects the reproductive success of grouse. **1**

d) (i) From **Figure 3**, describe what happens to the number of grouse as the percentage of the population treated by the drug is increased.
Give full explanations for this. **2**

(ii) Treating the entire population of grouse in the study with this drug is unlikely to ensure that the study area will be completely free of parasites in the following year.
Give **two** reasons for this. **2**

2 The dipper, *Cinclus cinclus*, is a small bird that lives by fast-flowing rocky streams in northern and western Britain. The birds feed on invertebrates, which they pick from beneath the water surface. They build their nests on ledges or under rocky overhangs on steep river banks. Dippers breed comparatively early in the year, but the date of egg-laying varies.

During 1987, observers in Wales monitored pairs of breeding dippers and recorded the date on which their first egg was laid. The pH of the water in the nest area was also measured. **Graph 1** below shows the results of their investigation. The invertebrates eaten by dippers in different pH condition was also observed. **Graph 2** shows how the percentages of stoneflies and mayflies in the diet was related to the pH of the water.

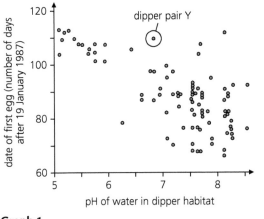
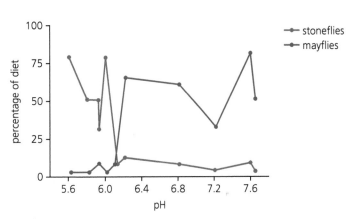

Graph 1   **Graph 2**

It was concluded that the less acidic the water in their habitat, the earlier in the year the birds' egg-laying date.

a) The hypothesis being tested in the investigation above is that water pH affects egg-laying date in dippers.
Suggest **two** other factors that might be affecting egg-laying date in this species. **2**

b) (i) The data in **Graph 1** can be attributed to correlation or causation.
From the data given, decide which you agree with and justify your choice. **1**

(ii) Compare the relationship between the pH of water and egg-laying date between pH 5.0–6.0 and pH 7.0–8.0. **2**

c) Suggest **one** factor related to the validity and **one** factor related to the reliability of the experimental design that would have to be taken into account when making conclusions from the results in **Graph 1**. **2**

d) From **Graph 2**, describe how the pH of water affects the percentages of stonefly and mayfly in diets of dippers. **2**

e) From **Graphs 1 and 2**, describe the diet of dipper pair Y. **2**

3 Feather mites of the order *Sarcoptiformes* are parasites of many birds. The mites feed on oil produced by the birds' oil glands. Oil is applied to feathers during preening times in the mornings and evenings, and empties the oil gland. Birds unable to oil their feathers efficiently use more energy maintaining body temperature. ⇨

A field investigation into the relationship between mite infestation and the breeding success of crested tits (*Parus cristatus*), a species given special protection in Scotland, was carried out. Crested tit nests were located using a systematic search method at a woodland study site and a number of those located were sampled at random. The parent birds from the sample were caught and the following measurements made.

- number of feather mites present
- size of the oil gland.

The nests were monitored and the following data collected:

- number of eggs laid
- number of chicks hatched
- number of chicks that survived to leave the nest.

The breeding success rate was calculated as the percentage of eggs laid from which chicks survived to leave the nest. The results are shown in the table and graph below.

Number of feather mites on parent birds	Breeding success rate (%)
0	86
2	100
5	64
10	82
14	70
15	85
170	42

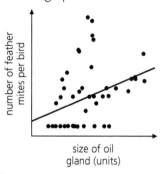

a) Suggest the reasons for the following:
   (i)   a systematic search was carried out to locate the crested tit nests **1**
   (ii) a random sample of the nests located was used in the study. **1**
b) Give a null hypothesis appropriate to the investigation. **1**
c) How does the data support the conclusion that feather mite infestations reduce breeding success in crested tits? **1**
d)(i) Describe the relationship between size of oil gland and number of feather mites per bird. **1**
   (ii) Suggest **one** precaution that should be taken to ensure that the oil gland measurements could be validly compared. **1**
e) Identify **one** precaution the investigators should take when working with protected species during their breeding cycle. **1**

Extended response

4 Give an account of the mean, mode and median of a data set, and describe the type of data for which median and mode might give better measures of central tendency than mean. **5**
5 Discuss correlations and how they are recognised. **5**

Answers can be found at pages 162–163.

Key Area 3.3

Reporting and critical evaluation of biological research

Key points !

1. Scientific reports should contain an explanatory **title**, an **abstract** including **aims** and findings, an **introduction** explaining the purpose and context of the study, and include the use of several **sources**, supporting statements, **citations** and **references**. ☐
2. Background information should be clear, relevant and unambiguous. ☐
3. The title should provide a succinct explanation of the study. ☐
4. The abstract should outline the aims and findings of the study. ☐
5. The aim must link the independent and dependent variables. ☐
6. The introduction should provide any information required to support choices of method, results and discussion; it should explain why the study has been carried out and place the study in the context of existing understanding. Key points should be summarised and supporting and contradictory information identified. ☐
7. Several sources should be selected to support statements, and citations and references should be in a standard form. Decisions regarding basic selection of study methods and organisms should be covered, as should the aims and hypotheses. ☐
8. The **method** section should contain sufficient information to allow another investigator to repeat the work. ☐
9. Experimental design should address the intended aim and test the **hypothesis**. ☐
10. The validity and reliability of the experimental design should be evaluated. An experimental design that does not address the intended aim or test the hypothesis is invalid. ☐
11. Treatment effects should be compared to controls. ☐
12. Any confounding variables should be taken into account or standardised across treatments. ☐
13. The validity of an experiment may be compromised when factors other than the independent variable influence the value of the dependent **variable**. ☐
14. **Selection bias** is the selection of a sample in a non-random way, so that the sample is not representative of the whole population. ☐
15. Sample size may not be sufficient to decide without bias whether the change to the independent variable has caused an effect in the **dependent variable**. ☐
16. The use of graphs, mean, median, mode, standard deviation and range for use in interpreting data should be appropriate to the **data collected**. ☐
17. In **results**, data should be presented in a clear, logical manner suitable **for analysis**. ☐
18. Consideration should be given to the validity of **outliers** and **anomalous results**. ☐
19. Statistical tests are used to determine whether the differences between the means are likely or unlikely to have occurred by chance. ☐
20. A statistically significant result is one that is unlikely to be due to **chance alone**. ☐
21. Error bars can indicate the variability of data around a mean or the percentage chance that the true mean lies somewhere along the bar. ☐
22. If a treatment mean differs from the control mean sufficiently for their error bars not to overlap, this indicates that the difference may be **significant**. ☐
23. Conclusions should refer to the aim, the results and the hypothesis. ☐
24. The validity and reliability of the experimental design should be taken **into account**. ☐
25. Consideration should be given as to whether the results can be attributed to correlation or causation. ☐

⇨

26 **Evaluation** of conclusions should also refer to existing knowledge and the results of other investigations. ☐

27 Meaningful scientific **discussion** would include consideration of findings in the context of existing knowledge and the results of other **investigations**. ☐

28 Scientific writing should reveal an awareness of the contribution of scientific research to increasing scientific knowledge, and to the social, economic and industrial life of the community. ☐

Reporting in the scientific cycle

Reporting of findings is part of the red area in the scientific cycle shown in Figure 3.14. It is done in an agreed, standard way that follows sets of rules designed to make reports easier to read and understand.

Writing a scientific report

There could be questions in your exam about these points, and understanding them is crucial for your project. Figure 3.15 outlines the main sections of a scientific report.

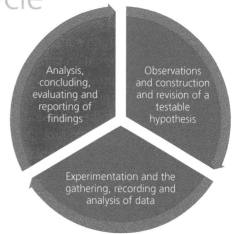

Figure 3.14 The reporting of findings area of the scientific cycle is shown in red

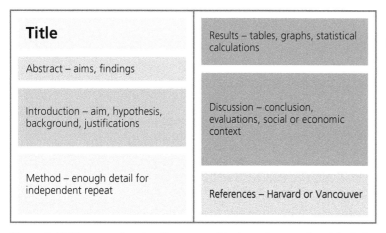

Figure 3.15 Diagram of a scientific report with the various sections labelled

Check-up 57

1 Give **three** types of information that should be present in the introduction to a scientific paper. **3**
2 Give the key feature of the method section of a scientific paper. **1**
3 Comment on the main characteristics of a results section of a scientific paper. **3**
4 Give **three** categories into which the contents of a discussion section of a scientific paper should fall. **3**
5 Give an account of the use of a lab book when carrying out a project/ research investigation. **5**

A scientific report must start with a meaningful title. This should be followed by the sections outlined in the table below.

Section	Notes
Abstract	A brief summary of the aims and main findings of the work.
Introduction	There should be an aim, which mentions the independent and dependent variables. Background information should set the work into context and describe its purpose. A hypothesis should be suggested. There should be reasons to justify the methods and procedures adopted where appropriate.
	Statements made here must be supported by citations to sources that have been published. Citations from sources not yet been published include personal communication from other scientists or papers that are still in press. All citations must be referenced by a standard method.

Section	Notes
Method	This should contain enough detail for the procedures to be repeated by an independent investigator. It should be appropriate for the aims and should be valid and reliable. If sampling is involved, the samples should be representative and be randomly taken to avoid selection bias.
Results	Clear, logical presentation of results data suitable for analysis. Validity of outliers and anomalous results should be considered. Graphs and statistical tests for use in interpreting data should be appropriate to the data collected. Graphs and tables should contain indications, such as error bars, of variability of data.
Discussion	Conclusions should refer to the aim, the hypothesis and the results data. Evaluation of experimental design should consider validity and reliability. Evaluation of conclusions should consider if the results suggest correlation or causation, and should refer to existing knowledge and the results of other investigations. Discussion should show awareness of the contribution of scientific research to new scientific knowledge, as well as its impact on the social, economic and industrial life of the community.
References	These should be presented using either the Harvard or Vancouver referencing systems.

Key words

Abstract – a brief outline of the aims and findings of an investigation

Aim – statement that links the independent and dependent variables

Anomalous result – result data that clearly does not fit in with the remainder of the data obtained; it may be discarded if sufficient doubt about it exists

Citations – a brief statement of the origin or source of a statement or of data

Discussion – consideration of the findings of an investigation, including placing results into context and against existing knowledge

Evaluation – consideration of methods and results leading to statements regarding possible errors and including suggested improvement

Introduction – an explanation of the background and context to an investigation

Method – details of the way an investigation has been carried out, including procedures used in experiments, with enough detail so that it could be repeated independently

Outliers – pieces of experimental data that lie well outside the basic range of data obtained

References – detailed information on where a citation has come from

Results – data sets obtained in an investigation and any processing carried out, including graphical presentations of data

Selection bias – error introduced when selection of samples has not been carried out in a random way

Sources – original origin of information for an investigation including other papers, reports, texts and internet sites

Title – meaningful heading for a scientific paper or report

Exam-style questions

Structured question

1 Give the meaning of the following terms used in experimental reporting:
   a) abstract   2
   b) citations   2
   c) outliers   2
   d) selection bias.   1

Extended response

2 Give an account of the main stages in the scientific cycle.   8

Answers are given on page 163.

Choosing your topic

You are required to research and report on a topic to demonstrate application of skills and knowledge in biology at a level appropriate to Advanced Higher in a context that is one or more of:

- unfamiliar to you
- familiar to you but investigated by you in greater depth
- integrates a number of contexts that are familiar to you.

The topic you settle on must be chosen with guidance from your teachers. It should allow you to collect numerical data for graphical analysis so that you can access all of the marks available.

Carrying through your project

Your project has two stages:

1   **Research stage**, in which you must plan and carry out experimental work and collect and analyse your experimental data. You must also gather information from books, journals and/or the internet to support your understanding of the biology underlying your project. It's crucial that your topic has a level of complexity that will allow you to access all of the marks.

During the research stage, you should keep a lab book to record your work and to form the basis of your report, although it is not actually assessed and is not submitted to SQA.

The research stage should take around 15 hours to complete.

2   **Report stage**, in which you must produce a written report on your research using your lab book as the basis. The report is submitted to SQA by a set submission date and must be your own individual work. It should be between 3000 and 3600 words in length (excluding the title page, contents page, tables of data, graphs, diagrams, calculations, references, acknowledgements and any appropriate appendices). You must supply a word count on the flyleaf of your report, and a penalty is applied to reports that exceed this count by more than 10%.

Marking instructions for the project report

It is strongly recommended that you read the published marking instructions for the project report, which are available to read or download on the SQA website: www.sqa.org.uk/files_ccc/ AHCATBiology.pdf

The table shows a summary of the total marks available for each category of the marking instructions along with some notes detailing where individual marks can be awarded. This table is not a substitute for the important information to be found in the published marking instructions, but it is a handy reference tool.

Section	Expected results	Marks
Abstract **1 mark**	A brief abstract stating main aim(s) and overall findings/conclusions	1
Introduction **5 marks**	Clear statement of aim(s) together with relevant hypotheses	1
	• Accounts of underlying biology relevant to aim(s) • Biological terms/ideas explained clearly and accurately • Biological terms/ideas at an appropriate depth • Biological importance justified	4
Procedures **9 marks**	Appropriate to aims(s)	1
	Procedures described clearly in sufficient detail to allow the investigation to be repeated	2
	Appropriate controls identified	1
	Control of confounding variables described	1
	Sample size appropriate	1
	Independent replication described and separate data set(s) provided	1
	Justification of how the pilot study informed the final procedures	1
	Shows complexity, creativity or accuracy	1
Results **6 marks**	Data relevant to the aim(s)	1
	Raw data recorded and within limits of accuracy of measurement	1
	Results presented appropriately	1
	Overall results calculated and presented appropriately	1
	Presentation of tables and graphs correct and accurate	2
Discussion (conclusion(s) and evaluation) **7 marks**	Conclusion(s) relevant to the aim(s) and supported by data in the report	1
	Conclusion(s) valid	1
	Evaluation of procedures with justification: • Means by which accurate measurements were achieved/sources of error in measurement and their impact on the results • Why the sample size was appropriate and how independent replication was achieved • How controls contribute to the overall validity of the investigation	2
	• How confounding variables were controlled or monitored and their impact on the validity of results • Solutions to problems and reasoning behind • Modifications to procedures in light of the pilot study	
	Results analysed and interpreted, and findings discussed critically and scientifically: • Analysis of results • Interpretation of results • Critical and scientific discussion of significance of finding(s)	3
Presentation **2 marks**	Appropriate structure, with informative title, contents page and page numbers	1
	References cited in the text and listed using Harvard or Vancouver referencing systems	1
Total		**30**

Understanding standards

We recommend that you have a look at the SQA Understanding Standards website: www. understandingstandards.org.uk

Here you will find a number of exemplar project reports, which are definitely worth looking through. The key support is to be found in the commentary that goes with each report and shows the overall marks for that project and details of why, or why not, marks have been awarded. This really is a must read!

Advice for project reports – avoiding common pitfalls!

Some of the areas that have been weaker in the project work in recent years are detailed in the table below – these are obviously worth bearing in mind when writing up your project. Take note of this advice, and consider reading through it before you submit your project. Use it as a checklist to confirm that you have taken in that piece of advice if it is appropriate.

Section	Advice	Tick if checked
Introduction	Ensure the underlying biology includes the necessary breadth and depth	
	Justifications for choices of method and procedures should be linked strongly to the actual investigation	
	Avoid presenting large amounts of information that is not clearly linked to the aim(s)	
	Avoid errors and inaccuracies in the biology presented	
	Ensure that sources used are of high quality	
	Address all of the biology that is fundamental to the topic being studied	
Method and procedures	Include all key information required to repeat the procedure	
	Attempt to control confounding variables made strongly	
	Avoid trying to justify a failure to control or monitor key confounding variables	
	Describe clearly how independent replication was achieved	
Results	Avoid basic errors with headings, scales, labels, units and plots	
	Be sure to summarise data by combining data sets	
	Support summarised data in graphs with appropriate tables	
Discussion	Avoid taking an overly simplistic approach and failing to show depth of understanding of the key issues affecting the validity and reliability of conclusions	
	Do not make invalid conclusions that do not accurately reflect the data presented	
	Failing to have appropriate controls can lose you marks in this section too	
	Inadequate repeat or replicate measurements can lose you marks in this section too	
	Evaluate procedures by providing more than a description of equipment and the possible errors associated with its use	
	Ensure the importance of controls is explained	
	Do not limit discussion of reliability to simple statements that the inclusion of repeats and replicates increased reliability	

Section	Advice	Tick if checked
Discussion	Identifying flaws that should have been addressed at the planning stage of the investigation, such as a failure to control key confounding variables, cannot regain you lost marks	
	Ensure that you stress adequately how pilot studies have affected your final experimental design	
	Be aware that the evaluation of your results is the most challenging part of the report	
	Be aware that investigations that are too simplistic offer limited scope for meaningful discussion	
	Be sure that you have discussed the variation in results between repeats and replicates	
	If you use statistical analysis, ensure that you have shown understanding of the statistical tools you have used	
	If you have calculated standard deviation or standard errors, ensure that you refer to them in your discussion	
	If you plot range bars, use them appropriately but don't necessarily conclude that differences between values were not significant if these bars overlap	
	Ensure that you relate your findings to relevant biology. Failure to include appropriate background information, including information from previous studies, in the underlying biology section can lose you marks here	
Presentation	Ensure that you cite and list a minimum of three references using either the Vancouver or Harvard system of referencing	
	If you use a website source, give required reference information, such as the organisation responsible for the material	
	Only giving the URL and date accessed is not acceptable	
	Identify journal articles clearly, even if they were accessed online; give the journal reference and not a URL	

Practice course assessment: investigative biology

Section 1

1 Which flowchart shows the sequence of events in the scientific cycle?
   A design experiment → form a new hypothesis → gather and analyse data → draw conclusions → form a hypothesis
   B form a hypothesis → design experiment → draw conclusions → gather and analyse data → form a new hypothesis
   C design experiment → gather and analyse data → form a hypothesis → draw conclusions → form a new hypothesis
   D form a new hypothesis → design experiment → gather and analyse data → draw conclusions → form a new hypothesis

2 The information in the table below explains terms used in biological science investigations.

   Which line in the table is **not** correct?

	Term	Explanation
A	Pilot study	Guides modification of experimental design
B	Hypothesis	Proposes an association between the independent and dependent variable
C	Confidence interval	Indicates the variability of the data around a mean
D	Positive control	Provides results in the absence of the treatment

3 An *in vitro* study involves observations made
   A in the natural habitat of an animal
   B in a living organism
   C in tissue cells growing in a culture medium
   D in extracts prepared from living tissues.

4 Turbidity measurements can be used to measure a population of micro-organisms indirectly. With a colorimeter, the higher the absorbance of light, the more yeast cells are present.
   Which row in the table below describes the variable being measured?

	Discrete	Continuous	Qualitative	Quantitative
A	✓		✓	
B	✓			✓
C		✓	✓	
D		✓		✓

5 Which of the following describes the purpose of a randomised block design?
   A Reduce effect of confounding variables
   B Ranking sample data
   C Ensuring that sampling is representative
   D Controlling the independent variable

Section 2

1 Adiponectin is a signalling molecule thought to increase the sensitivity of cells to insulin. The table shows the results of a clinical study in which increases in adiponectin concentration following treatments received by individuals at risk of developing type 2 diabetes was determined.

Treatment	Average increase in concentration of adiponectin in blood plasma ($\mu g\ cm^{-3}$)
Drug treatment	0.83 ± 0.05
Lifestyle changes	0.23 ± 0.05
Control (no treatment)	0.10 ± 0.05

a) Compare the results of drug treatment to lifestyle changes in terms of their effectiveness in increasing adiponectin concentration. **1**

b) The clinical study used human volunteers.

   (i) Give **one** ethical issue that should be considered when using human volunteers. **1**

   (ii) Explain why large numbers of volunteers are required to produce reliable results upon which valid conclusions may be based. **1**

2 In spring, male 15-spined sticklebacks build a nest and then attempt to attract females to lay eggs in it. Only the male looks after the young, 'fanning' the nest to increase the flow of oxygenated water through it. Female choice of males seems to be based on the behaviour of the male after he has built his nest, particularly the time spent fanning, which begins as soon as the nest is completed. The scatter plot below shows the relationship between the number of fanning bouts per hour and hatching success.

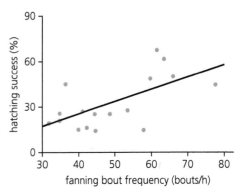

a) Identify the independent and dependent variables in this study. **2**

b) Explain the relationship shown in the scatter plot. **2**

c) The charts below compare the fanning behaviour of males that had a low or high hatching success.

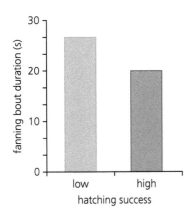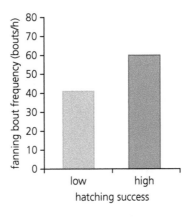

Discuss the best strategy for males to adopt in order to increase the hatching success of their eggs. **2**

3 The scatter plot below shows the results of a nine-year study in an area of northwest Russia. Each year scat samples from pine martens were analysed for the remains of species of grouse, and were plotted against the relative abundance of voles, a small mammal often eaten by martens. Each plot represents a year; the 1958 plot was treated as an outlier for the purpose of drawing overall conclusions from the data.

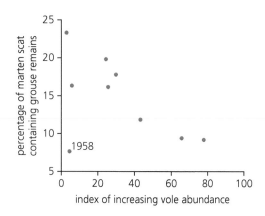

a) Describe what is meant by a scat sample. **1**

b) (i) Describe the correlation shown by the scatter plot. **1**

(ii) It was concluded that the numbers of grouse in pine marten diets increased in years of low vole abundance.

Suggest a possible confounding variable that could invalidate this conclusion. **1**

(iii) Explain what is meant by an outlier. **1**

(iv) Suggest what could be done to confirm that the outlier was an anomalous result. **1**

4 As part of a study on the breeding density of blue and great tit pairs in a woodland, an increasing number of nest boxes were erected.

Graphs 1a and 1b show how the density of pairs of blue tits and great tits changed as the number of nest boxes erected in eight different sample areas in the woodland increased.

Graph 2 shows the relationship between nest box density in the eight sample areas and the percentages of blue tit pairs among the total number of both blue and great tit pairs using boxes in the same woodland area.

Graph 1 a) and b)

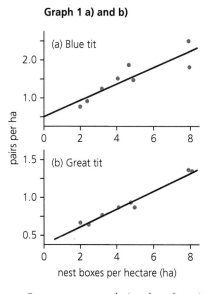

Graph 2

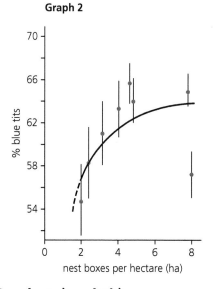

a) Compare trends in the data in **Graphs 1a) and 1b)**. **2**

b) (i) Describe the general trend shown by **Graph 2**. **1**

(ii) Comment on the information given by the error bars in **Graph 2**. **1**

c) A piece of woodland in the study area was 125 hectares and had 25 nest boxes erected within in it. Use the data above to calculate the percentage of these boxes expected to be occupied by great tit pairs. Show all of your working **2**

5 a) Discuss the advantage of a pilot study in the development of a biological investigation. **5**

OR

b) Give an account of scientific ethics in the use of animals and humans in drug trials. **5**

Answers can be found on pages 163–164.

Advanced Higher exam advice for biology

You should have a copy of the course specification (CS) for Advanced Higher Biology but, if you haven't got one, download it from the SQA website (www.sqa.org.uk). It is worth spending some time looking at this document, as it indicates what you can be tested on in your examination.

This book contains check-up questions and exam-style questions that have been carefully assembled to be as similar as possible to typical Advanced Higher Biology examination questions. The marking instructions give acceptable answers with alternatives. You should also be attempting the past papers, which are on the SQA website, and using their answers to mark your own work.

The following hints and tips are related to examination techniques, as well as avoiding common mistakes. Remember that if you hit problems with a key area or a specific question, you should ask your teacher for help.

Section 1

- 20 multiple-choice items worth 20 marks in total.
- Answer on a grid.
- Do not spend more than 30 minutes on this section.
- Some individual questions might take longer to answer than others – this is quite normal. Make sure you use scrap paper if a calculation or any working is needed.
- Some questions can be answered instantly – again, this is normal.
- Do not leave blanks – complete the grid for each question as you work through.
- Try to answer each question in your head without looking at the options. If your answer is there, you are home and dry!
- If you are not certain, it is sometimes best to choose the answer that seemed most attractive on first reading the answer options.
- If you are guessing, try to eliminate options before making your guess. If you can eliminate three, you will be left with the correct answer even if you do not recognise it.

Section 2

- Restricted and extended response questions worth 80 marks in total.
- Spend about two hours on this section.
- A clue to your answer length is the mark allocation – questions restricted to 1 mark can be quite short. If there are 2–3 marks available, your answer will need to be extended and may well have two, three or even four parts.
- The questions are usually laid out in Area sequence, but remember that some questions are designed to cover more than one Area.
- C-type questions that start with 'State', 'Identify', 'Give' or 'Name' often need only a single sentence in response. They will usually be for 1 mark each.
- Questions that begin with 'Explain', 'Suggest' and 'Describe' are usually A-type questions and are likely to have more than one part to the full answer. You will usually have to write a sentence or two, and there may be 2 or 3 marks available.

- Make sure you read over the question twice **before** trying to answer – there will be very important information within the question, and underlining or highlighting key words is good technique.
- Using abbreviations such as PAGE and Rb is fine. The Advanced Higher Biology course specification will give you the acceptable abbreviations.
- Don't worry if the questions are in unfamiliar contexts, that's the idea! Just keep calm and read the questions carefully.
- In the large data question (Q1, which will be worth between 7 and 10 marks), it is good technique to read the whole stem and then skim the data groups before starting to answer any of the parts.
- In the large data question (Q1), be aware that the first piece of data presented should give the main theme of the question.
- In the large experimental design question (which will be worth between 5 and 9 marks), you must be aware of the different classes of variables, why controls are needed, and how reliability and validity might be improved. It is worth spending time on these ideas – they are essential and will come up year after year.
- Note that information that is additional to the main stem may be given within a question part – if it's there, you will need it!
- If instructions in the question ask you to refer to specific groups of data, follow these and don't go beyond them.
- Remember that a conclusion can be seen from data, whereas an explanation will usually require you to supply some background knowledge as well.
- Note that, in your answer, you may be asked to 'use data to …' – it is essential that you do this.
- Remember to 'use values from the graph' when describing graphical information in words, if you are asked to do so.
- Look out for graphs with two y axes – these need extra concentration and anyone can make a mistake.
- In numerical answers, it's good technique to show your working and supply units.
- Answers to calculations will not usually have more than two decimal places.
- You should round any numerical answers as appropriate, but two decimal places should be acceptable.
- Ensure that you take error bars into account when evaluating the effects of treatments.
- There will be two extended response questions for a total of between 12 and 15 marks. The longer question will contain a choice – ensure you spend a bit of time making your choice.
- Do not leave blanks. Always have a go, using the language in the question if you can.

Answers

Area 1 Cells and proteins answers

1.1 Laboratory techniques for biologists

Answers to check-up questions

Check-up 1

1 Toxic chemicals, corrosive chemicals, flammable chemicals, pathogens, mechanical hazard. **5**
2 Identifying risks; identifying control measures. **2**

Check-up 2

1 Mix 9 ml of stock with 1 ml solvent to give 0.9 M, 8 ml stock with 2 ml solvent to give 0.8 M, and so on. **3**
2 Carry out serial log dilutions to give 10^{-5}/very dilute suspension; plate out dilute suspension and incubate; count colonies; calculate original cell density. **4**

Check-up 3

1 Make up glucose solutions of known concentration; put each through colorimeter to determine absorbance; draw standard curve; put unknown through colorimeter and determine absorbance; read concentration from standard curve. **Any 4**
2 Solutions that resist pH changes; used to maintain pH level in experiments in which pH is a possible (confounding) variable. **2**

Check-up 4

1 Centrifugation separates by density; the most dense material is in the pellet, the least is left in the supernatant. **2**
2 Add spot of solution to origin line of a paper/thin layer chromatography strip; dip in (appropriate) solvent and allow it to run; develop with amino acid stain. **3**
3 Add mixture to column; target protein binds to ligand/antibody in column; other proteins run through; target protein then collected by washing out the column. **Any 3**

Check-up 5

1 In native PAGE, proteins are separated in electric field on the basis of mass and charge; in SDS–PAGE, proteins are denatured/all given negative change and separated in electric field on the basis of mass. **2**
2 At its isoelectric point, a protein has no net charge; it precipitates; if a mixture is run through pH gradient gel, each protein precipitates at its isoelectric point. **3**

Check-up 6

1 Supply of antibodies all with the same specificity/which bind to the same antigen. **1**
2 Bound to an antibody that binds to an antigen in immunoassay well; substrate added, which is converted to coloured product. **3**
3 Proteins separated by gel electrophoresis; separated protein blotted on to membrane; treated with antibody with reporter; reporter visualised/example. **4**
4 To prevent contamination; prevent competition with target microbe. **2**
5 38 000 viable cells; 10 000 dead cells. **2**

Answers to exam-style questions

Structured questions

1 a) 200 000 cells per cm^3 **1**
   b) It stains live cells only. **1**
   c) Suitable nutrients *or* energy source/glucose/amino acids/fatty acids/vitamins; serum/growth factors. **2**
   d) It prevents entry of contaminants; ensures cultures are pure. **2**
2 a) (i) 97.9% **1**
        (ii) Seven times **1**
   b) When proteins (in solution) are brought to their isoelectric point, they have an overall neutral charge and precipitate out of solution. **2**
   c) Enzyme binds to ligand/antibody and becomes trapped in the column; column

can then be washed out and the (purified) enzyme released. **2**

Extended response

3 Amino acids separated by paper/thin layer chromatography; based on solubility/affinity; dependent on R group; protein separated by gel electrophoresis based on mass/charge; or by isoelectric point; based on pH at which they have no net charge/are neutral; and precipitate from solution. **7**

4 To identify antigen; using monoclonal antibodies; bind specifically to antigens; antibodies carry reporter enzymes/tags/labels; which can be visualised; by colour change/fluorescence/radioactivity; only when antigen present (in assay well). **7**

1.2 Proteins

Answers to check-up questions

Check-up 7

1 The entire set of proteins expressed by the genome. **1**

2 An individual gene may be expressed to produce several different proteins because of alternative splicing. **1**

3 (During splicing) different exons may be retained; different introns may be removed. **2**

4 Metabolic activity, which changes with age/senescence/dormancy state; cellular stress level, which depends on extremes of temperature/pH/exposure to toxins/mechanical damage; response to signalling molecules such as hormones/antigens exposure; state of health; apoptosis. **Any 3**

Check-up 8

1 Endoplasmic reticulum (ER); is a system of membrane tubules; continuous with nuclear membrane; can be rough (RER) with ribosomes; or smooth (SER) without ribosomes; Golgi apparatus is related to the ER; forms vesicles/involved in the formation of lysosomes. **Any 4**

2 Phospholipids synthesised in SER; transmembrane proteins started in cytosolic ribosomes; finished in docked ribosomes/RER. **3**

Check-up 9

1 Cytosolic proteins are synthesised in cytosolic ribosomes; transmembrane proteins start in cytosolic ribosomes and finish in docked ribosomes/RER; inserted into ER membrane; secretory protein started in cytosolic ribosomes and finished in docked ribosomes/RER; released into ER lumen. **Any 4**

2 Addition of carbohydrate in Golgi apparatus; proteolytic cleavage of digestive enzymes in gut. **2**

Check-up 10

1 Hydrophobic are uncharged; basic are positively charged/hydrophilic; acidic are negatively changed/hydrophilic; polar have small balanced charges/have no (net) charge. **4**

2 Primary is sequence of amino acids; secondary is folding to produce helices/sheets/turns; held in shape by hydrogen/H bonds. **3**

Check-up 11

1 Hydrophobic attractions; ionic bonds; covalent disulfide bridges; London dispersion forces; hydrogen/H bonds. **5**

2 Arrangement of subunits; may have prosthetic groups. **2**

3 Binding or dis-binding of substances at subunits; affects affinity of these substances for further subunits. **2**

Check-up 12

1 Binding of oxygen to first subunit; increases affinity of further subunits for oxygen *or* converse. **2**

2 Allosteric sites are binding sites but not the active site; positive modulators increase activity at binding site/are activators; negative modulators decrease activity at active site/are inhibitors. **3**

3 Phosphorylation is by kinases/changes conformation of protein/can activate protein; de-phosphorylation is by phosphatases/changes conformation of protein and can deactivate protein. **2**

Answers to exam-style questions

Structured questions

1. a) Starts in cytosolic ribosomes; finished in docked ribosomes/ribosomes on ER.   **2**
   b) From ER.   **1**
   c) Addition of carbohydrate.   **1**
   d) Vesicles fuse with cell membrane and contents are released.   **1**
   e) Proteolytic cleavage.   **1**
2. a) Alpha helix.   **1**
   b) That the molecule has subunits; four subunits in haemoglobin.   **2**
   c) Prosthetic group.   **1**
3. a) After binding to the first subunit; affinity with other subunit increases; and the oxygen loads faster.   **3**
   b)
   (i) Carbon dioxide; lactate/lactic acid.   **2**
   (ii) Respiring muscle needs oxygen; heat and acidity make oxygen bind less tightly to haemoglobin; easier to release to muscle cells.   **2**
   (iii) Higher temperature.   **1**

Extended response

4. Addition or removal of phosphate causes conformational changes in protein; form of post-translational modification; kinases add a phosphate group; phosphatases remove phosphate group; conformation changes affect protein activity; activity can be regulated in this way; some proteins are activated by phosphorylation, others are inhibited; phosphate group adds negative charge, which can change ionic interactions in the protein.   **Any 5**
5. Classes depend on R groups: hydrophobic R group has no charge; hydrophobic attractions; London dispersion forces between hydrophobic groups; basic R group has a positive charge; form ionic bonds; acidic R group has a negative charge; form ionic bonds; polar R group balances tiny charges; form hydrogen bonds; sulfur-containing amino acids can form covalent bonds/disulfide bridges.   **Any 8**

1.3 Membrane proteins

Answers to check-up questions

Check-up 13
1. Phospholipid; intrinsic proteins/transmembrane proteins; peripheral proteins.   **3**
2. Transmembrane bound by hydrophobic interactions; hydrophilic bound on surfaces; peripheral bound to transmembrane proteins/phospholipid.   **3**

Check-up 14
1. Act as channels; transporters.   **2**

Check-up 15
1. Ligand-gated channels have a receptor; specific to a ligand/signal molecule/hormone/neurotransmitter; open when ligand bound; voltage-gated channels open or close in response to changes in membrane potential.   **4**
2. ATP dephosphorylated by transporter protein; energy released; to move substances against their concentration gradient.   **3**

Check-up 16
1. Binds three sodium ions from inside cell; phosphorylated by ATP and changes conformation; releases sodium outside and binds two potassium ions from outside; dephosphorylated and changes conformation; releases potassium on inside.   **Any 4**
2. Sodium moves through symport down its concentration gradient; energy released; and used to move glucose in the same direction (even against its concentration gradient).   **3**

Answers to exam-style questions

Structured questions

1. a) Oxygen/carbon dioxide/hydrophobic signal molecule.   **Any 1**
   b) State 1 – low affinity; state 2 – high affinity.   **2**

c) Two potassium in to three sodium out. **2**

2 a) R, S, T integral; V peripheral. **2**

b) Hydrophobic attractions. **1**

c) Channel protein; forms a pore; allows facilitated diffusion of substances (in and out of the cell). **3**

Extended response

3 Ligand-gated channels; have receptors specific to signal molecules; binding opens channels. **3**
Voltage-gated channels; membrane potential/voltage changes open/close channels. **2**

4 Transports glucose and sodium together/simultaneously; into intestinal cells from intestine lumen; sodium moves down its concentration gradient; providing energy for glucose movement; against its concentration gradient; sodium–potassium pump maintains sodium gradient/low concentration of sodium in intestinal cells. **Any 5**

1.4 Communication and signalling

Answers to check-up questions

Check-up 17

1 Pass directly through phospholipid bilayer. **1**

2 Enter cell and bind with receptor in cytosol/nucleus; hormone–receptor complex binds to DNA in nucleus; at hormone response elements (HRE); affect transcription of genes. **Any 3**

Check-up 18

1 Signal molecule/hormone acts as a ligand; binds to complementary/specific receptor in cell membrane. **2**

2 When signal bound, receptor causes G-protein/kinase cascade; activates cellular response. **2**

Check-up 19

1 Insulin binds to receptor; phosphorylation cascade; GLUT4 vesicles recruited into membrane; GLUT4s allow glucose into cell. **4**

2 Type 1 from birth; lack of insulin; type 2 linked

to obesity; receptors are not sensitive to insulin signal. **4**

Check-up 20

1 Neurotransmitter binds to receptor; sodium channels open; change in potential opens voltage-gated sodium channels; further sodium entry takes membrane to threshold. **4**

2 Sodium channels close; potassium channels open and potassium ions leave; sodium–potassium pump is activated; pump causes net loss of sodium ions. **4**

Check-up 21

1 Opsin; bound to retinal. **2**

2 Photon affects conformation of rhodopsin to excited state; cascade of hundreds of transducins; activate PDE; PDE breaks down cyclic GMP/cGMP; reduction in cGMP closes ion channels and impulse fires. **Any 4**

Answers to exam-style questions

Structured questions

1 a) Retinal and opsin. **2**

b) They activate a cascade of phosphodiesterase (PDE) molecules; which breakdown cyclic GMP/cGMP molecules. **2**

c) Sodium ion channels close. **1**

d) It allows vision in dim light/allows nocturnal/crepuscular activities/hunting/foraging. **1**

e) Different opsins. **1**

f) Allows colour vision. **1**

2 a) Opens sodium ion channels; increases positive change inside neuron membrane. **2**

b) Increased positive change inside membrane opens voltage-gated channels; allows more sodium ions to enter. **2**

c) The potential at which the neuron will transmit the impulse. **1**

d) Potassium channels open/sodium channels close; sodium–potassium pump restores membrane potential. **2**

e) Allows the system/neuron to remain sensitive to further stimulation. **1**

Extended response

3 Insulin binds with receptor; triggers a cascade of phosphorylation; GLUT-containing vesicles are recruited into membrane; glucose enters through the GLUT from bloodstream. **4**

4 Hydrophobic signal/steroid hormone; example of steroid hormone; enters cell through phospholipid layer; combines with receptor in cytosol/nucleus; hormone–receptor complex binds with hormone response elements (HRE) on DNA; gene expression affected. **6**

1.5 Protein control of cell division

Answers to check-up questions

Check-up 22

1 Gives cell support and shape; polymerisation and depolymerisation of tubulin. **2**

2 M phase divides by mitosis; G1 growth; S (synthesis) DNA replicates; G2 further growth. **4**

3 Prophase: DNA condenses into chromosomes; Metaphase: chromosomes move to metaphase plate/equator and attach to spindle by kinetochores; Anaphase: sister chromatids separated and chromosomes move towards opposite poles; Telophase: chromosomes start to decondense. **4**

Check-up 23

1 Cyclins accumulate; CDKs phosphorylate Rb; Rb inhibited so genes transcribed; to allow proteins needed for S to be synthesised. **4**

2 G2 checkpoint: DNA replication/damage assessed; go-ahead allows progress to M phase. **2**
Metaphase checkpoint: assess chromosome attachment to spindle; go-ahead allows mitosis to complete. **2**

3 A proto-oncogene mutates to form a tumour-promoting oncogene. **1**

Check-up 24

1 By external death signals/lymphocytes; internal death signals/activation of p53; lack of cell growth factors. **3**

2 Allows metamorphosis in some species; can lead to tumour suppression/destruction. **2**

Answers to exam-style questions

Structured questions

1 a) (i) Uncontrolled decrease – degenerative disease; uncontrolled increase – tumour formation. **2**
    (ii) Do not receive go-ahead at G1; because not enough cyclin produced/cyclin-dependent kinases not activated. **2**
    (iii) DNA is replicated. **1**
    (iv) Tubulin; is polymerised/depolymerised. **2**
    (v) Cytoplasm splits and cell divides into two. **1**

  b) (i) Cell grows. **1**
    (ii) Unphosphorylated Rb prevents S phase going ahead; cell enters G0/no S phase means no DNA replication; no S phase/cell in G0 prevents cell reaching M phase so no cell division and no tumour. **3**

Extended response

2 Internal death signal; activated p53 protein; causes caspase cascade; caspases break down cell structures causing cell death; external cell death signal; for example lymphocyte/natural killer cell; binds to receptor; causes caspase cascade. **Any 5**

3 Prophase; chromosomes/DNA condense; two sister chromatids visible; nuclear membrane breaks down; spindle microtubules extend from the MTOC by polymerisation; and attach to chromosomes via their kinetochores in the centromere region. **Any 4**
Metaphase; chromosomes align at the metaphase plate/equator; anaphase; microtubules shorten by depolymerisation; sister chromatids separated; and the chromosomes formed are pulled to opposite poles; telophase; the chromosomes decondense; two new nuclear membranes are formed. **Any 6**

Answers to practice course assessment

Section 1

1 C	3 D	5 C	7 B	9 B
2 C	4 C	6 A	8 D	10 A

Section 2

1 a) Immunoassay. **1**
   b) Antibodies derived/made from identical immune cells/the same unique parent immune cell. **1**
   c) The test is specific for HSV *or* the HSV antibody is specific to the HSV antigen *or* the antibody linked to the reporter enzyme is specific to the HSV antibodies. **1**
   d) Antibody R might be left in the container even though it is not attached, so the reporter enzyme might give coloured product/positive result. **1**
   e) pH influences reaction rates/activity of enzymes. **1**
   f) Radioactive/fluorescent. **1**

2 a) Sequence of amino acids. **1**
   b) Alpha helix; turns. **2**
   c) Primary based on strong/covalent bond *and* secondary on weak/H bonds. **1**
   d) (i) Prosthetic group. **1**
       (ii) Binding of oxygen alters conformation to increase affinity for more oxygen. **1**

3 a) Ligand/signal molecule binds to gate protein to change conformation; changes in ion concentration/potential across membrane changes conformation of protein. **2**
   b) Release energy from ATP; for use in active transport. **2**
   c) (i) Q, P, S, R **1**
       (ii) Nerve transmission/active transport of glucose **or** glucose symport. **1**

4 a) Binds to ligand-gated channel/receptor and signal induced/G-protein cascade **or** phosphorylation cascade; GLUT4 **or** glucose transporter proteins recruited to membrane and glucose enters cell. **2**
   b) Lower adiponectin; less sensitivity to insulin. **2**
   c) Exercise triggers/encourages GLUT4 recruitment into membranes; more glucose can be removed from blood **or** improves uptake of glucose to fat and muscle cells. **2**

5 a) Cells may be harmful/tumour forming/cancerous. **1**
   b) (i) Caspases. **1**
       (ii) Bcl-2 no longer made; apoptosis no longer occurs and so tumours can grow. **2**
   c) (i) DNA damage; activation of a tumour suppressor. **1**
       (ii) p53 **1**

6 a) (i) Synthesis of membrane components: lipids synthesised in the SER; are inserted into its membrane; synthesis of proteins begins in cytosolic ribosomes; synthesis of cytosolic proteins completed there; synthesis of transmembrane proteins begins in cytosolic ribosomes; signal sequence; ribosomes dock with the RER; transmembrane permanently attached there. **Any 6**
       (ii) Movement of proteins for secretion: Proteins for secretion, such as peptide hormones/digestive enzymes, or those which will become lysosome hydrolases; translated in ribosomes docked on the RER; enter its lumen; proteins move through the Golgi apparatus; packaged inside secretory vesicles; vesicles take proteins to the plasma membrane for secretion; some develop into lysosomes. **Any 4**
   b) In rod cells retinal-opsin complex is rhodopsin; retinal absorbs a photon of light; changes conformation to photoexcited rhodopsin; activates hundreds of G-protein molecules; called transducins; each of which activate one molecule of the enzyme phosphodiesterase (PDE); PDE catalyses the hydrolysis (breakdown) of a molecule called cyclic GMP (cGMP); each active PDE molecule breaks down thousands of cGMP molecules; reduction in cGMP concentration causes ion channels in the membrane of rod cells to close; triggers nerve impulses in neurons in the retina; high degree of amplification results; rod cells respond to low intensities of light. **Any 10**

Area 2 Organisms and evolution answers

2.1 Field techniques for biologists

Answers to check-up questions

Check-up 25

1 A hazard is a harmful side-effect of apparatus or a procedure; a risk is the likelihood of the hazard occurring. **2**

2 (Use of) mountain equipment/transport; first aid kit; appropriate clothing; appropriate footwear; means of communication; map/compass. **Any 5**

Check-up 26

1 Laws that protect environments/habitats/species. **1**

2 Observe rather than catch; minimise numbers sampled; return samples to habitats as soon as possible; spend minimum time in habitats. **Any 2**

3 A line transect is a single line with species touching the line at stations counted in; a belt transect is a wider zone with quadrats used at stations. **2**

4 A square frame of known area/side; used to sample plants/sessile/slow-moving species. **2**

5 A species that is difficult to observe; rare/inaccessible habitat/nocturnal/skulking; use remote detection/trail camera/camera trap. **3**

Check-up 27

1 To identify/name species; by using a series of paired statements/questions. **2**

2 The classification and naming of living organisms. **1**

3 The study of relationships between organisms; based on DNA sequences. **2**

4 Divergent based on same structures adapted for different functions; convergent based on different structures adapted for a similar function. **2**

5 Arthropoda – invertebrates with joined legs; Nematoda – round worms; Chordata – spinal cord. **3**

Check-up 28

1 A species whose presence/absence/abundance gives information about environmental quality. **1**

2 Latency – time between stimulus and response; frequency – number of repeats of a behaviour per unit of time; duration – time a behaviour is continued for. **3**

3 Ethogram provides the list of behaviours; observation allows durations of each behaviour (on list) to be recorded. **2**

4 Allocation of human characteristics/emotions to animal behaviour; could lead to misinterpretation/invalid conclusions. **2**

Answers to exam-style questions

Structured questions

1 a) (i) 12 694 **1**
     (ii) 1250 **1**
  b) (i) Frame quadrat **or** systematic sampling. **1**
     (ii) Paint/varnish mark on shell; consider effect on survival rate/breeding/predation/feeding of snail. **2**

2 a) Ethogram **1**
  b) Time between a stimulus and a response. **1**
  c) Anthropomorphism; could lead to misinterpretation/invalid conclusions. **2**
  d) Remote observation using camera traps/videoing the observations. **1**

Extended response

3 Point counts – counts from a fixed spot; transects – lines across habitat; quadrats – standard sample frames; netting/trapping – for mobile species; remote detection/camera traps/scat sampling – for elusive species. **Any 5**

4 Divergence is adaptation of the same structure for different function; so organisms look different but are closely related; convergence is the adaptation of different structures to the same function; so organisms look similar but are only distantly related. **4**

5 a) Model species are easy to study; information obtained can be used to infer/predict information; about related species/species in the same taxonomic group; which are difficult to study. **Any 3**

   b) Indicator species presence/absence/abundance; can give information about environmental factors; such as pollutants/habitat classification; indicator can be favoured/susceptible to factor. **Any 3**

2.2 Evolution

Answers to check-up questions

Check-up 29

1 Drift is random; caused by population bottlenecks/significant population reductions **or** founder effects/small sample of a population colonising a new area; in small populations. **2**

2 Can be deleterious; neutral; or advantageous. **3**

3 Intensity of a factor such as competition/predation/disease/temperature/humidity; that affects survival. **2**

4 Female choice; male–male rivalry. **2**

Check-up 30

1 No selection; no mutation; no migration; free interbreeding; large population. **5**

2 $p^2 + q^2 + 2pq = 1$ **1**

3 That evolutionary/selection processes were occurring. **1**

Check-up 31

1 Measure of ability of phenotype/genotype to survive/reproduce; can be absolute or relative. **2**

2 Ratio of the frequency of a genotype in a generation; to its frequency in the previous one. **2**

3 Ratio of number of surviving offspring of one genotype; to the most successful genotype. **2**

Check-up 32

1 Intimate; co-evolved relationship; between two species. **3**

2 Parasitism +/−; commensalism +/0; mutualism +/+. **3**

3 Parasites evolve greater virulence; which acts as selection pressure on hosts; which evolve greater resistance/tolerance; which acts as selection pressure on the parasite. **4**

Answers to exam-style questions

Structured questions

1 a) Species respond to selection pressures imposed by each other. **1**

   b) *Cyanea* +/birds + **1**

   c) *Drepanis pacifica* can feed without competition; because the flower tube of *Cyanea superba* is so long that only its beak can reach the nectar/food. **2**

   d) *Cyanea* plants evolved longer flower tubes to attract birds to feed; and so pollinate them; which acted as selection pressure on birds to evolve longer beaks to reach the nectar; which acted as selection pressure on the plants in case the birds could feed without pollinating. **Any 3**

2 a) p is frequency of B, and q is frequency of b; peppered wing moths are bb and there are 40% of these, so $q^2 = 0.4$
So, $q = \sqrt{0.4} = 0.63$
Since $p + q = 1$, then $p = 1 − 0.63 = 0.37$
So heterozygous moths = $2pq = 2(0.37 \times 0.63) = 0.47$ i.e. 47%
And homozygous dominant moths = $p^2 = (0.37)^2 = 0.14$

   (i) 47% **1**

   (ii) 0.14 **1**

   b) No natural selection; random mating; no mutation; no gene flow by migration; large population size. **Any 3**

Extended response

3 a) Mutation is random change to genetic information; can be deleterious/harmful or neutral or beneficial/advantageous. **Any 2**

   b) Genetic drift is random; affects small populations; due to bottleneck effects/description; or due to founder effects/description. **Any 3**

   c) Natural selection is non-random; it is about survival of fittest; favours characteristics that improve survival; causes increases in these characteristics in subsequent generations; sexual selection is about ability to breed;

favours characteristics that improve breeding success; can lead to sexual dimorphism. **Any 5**

4 a) Parasitism is a symbiosis in which the parasite benefits in terms of food/energy/resources/+; and the host is harmed by loss of resources/−; it is co-evolved between two different species. **3**

   b) Mutualism is a symbiosis in which both species gain/+/+; commensalism is a symbiosis in which one species gains/+/0; and the other neither gains nor loses/0; they are co-evolved. **4**

2.3 Variation and sexual reproduction

Answers to check-up questions

Check-up 33

1 Costs – only half population reproduce; breaks up successful genotypes; benefits – produces variation upon which selection can act. **3**

2 Parasites and hosts need to be variable; to be able to evolve quickly; changes in one partner require changes in the other; new adaptations in one partner put selection pressure on to the other partner. **4**

3 Conserves fit/successful genotypes; allows rapid reproduction rates. **2**

4 Development of egg without fertilisation; in animals; more common in colder climates/where parasites less common. **3**

5 Genes passed between individuals in a generation; allows acquisition of genes within the lifetime; produces rapid rates of evolution. **3**

Check-up 34

1 Same size and centromere position; same genes at same loci. **2**

2 Breaks up and recombines linked alleles/genes; producing varied recombinations. **2**

3 Each member of a homologous pair lines up independently; separates randomly regardless of origin. **2**

4 Chromosomes appear and condense; form homologous pairs on spindle; chiasmata form/recombination occurs; pair forms/separates independently of others; two haploid cells formed. **5**

5 Sister chromatids separate; four haploid gametes formed. **2**

Check-up 35

1 Sex chromosomes; two Xs in females and XY in males. **2**

2 Encodes testes-determining factor (TDF). **1**

3 Produce two different gametes either with X or Y chromosome. **1**

4 Sex-linked alleles on X chromosome and not on Y; carrier females have two alleles for characteristics, one on each X; males inherit only one allele of each sex-linked gene. **3**

5 Prevents double doses of any X allele product; female Xs deactivated randomly; only 50% of cells express the deleterious allele product. **3**

Check-up 36

1 Temperature; competition; infection; body size. **4**

2 Increased competition causes production of more males (in mouse lemurs); males move away, reducing competition for their mother. **2**

3 Has both male and female gametes; can mate with any other individual it meets. **2**

Answers to exam-style questions

Structured questions

1 a) (i)   Presence of a Y chromosome with the SRY allele. **1**

      (ii) Temperature/competition/infection/body size. **Any 2**

   b) (i)   Males only have one sex-linked allele for each gene; if it is deleterious, they will show the condition; females have two sex-linked alleles for each gene so they can be carriers. **Any 2**

      (ii) X chromosome deactivation means that in half of cells the deleterious allele is not expressed/no product is produced. **1**

      (iii) 50%, or one in two. **1**

2 a) Vertical is from parent to offspring; horizontal is from individual to individual/within a generation. **1**

   b) Rapid/fast colonisation of disturbed habitat; +/0 conserves successful genomes; mating not required. **Any 2**

c) (i) Mutation can still occur; some mutations may be advantageous and be selected for. **2**

(ii) Parthenogenesis **1**

(iii) Hermaphrodite **1**

Extended response

3 Homologous chromosomes pair in meiosis I; pairs assort independently; separation combines members of each pair randomly/regardless of maternal or paternal origin; chiasmata occur between non-sister chromatids in a homologous pair; leading to crossing over; and recombination of linked alleles; final variation depends on separation of sister chromatids in meiosis II. **All 7**

4 Variation needed for evolution; evolution needed for hosts to adapt to increasing virulence of parasites; adaptations allow resistance to/tolerance of parasite; tolerance requires new adaptations for virulence in the parasite; sexual reproduction produces the required variation; this benefit outweighs cost of sexual reproduction. **All 6**

2.4 Sex and behaviour

Answers to check-up questions

Check-up 37

1 Females invest more energy/resources into each egg structure (compared to each sperm in males); and into the uterus and gestation period in mammals (compared to males). **2**

2 Requires energy and resources to be consumed; enhances survival/production rate of young. **2**

3 r-selection – small body size/lifespan; reproduce early in life but less often; larger number of smaller offspring; K-selection – large body size/long-lived; offspring mature slowly; fewer but larger offspring. **Any 4**

4 Internal costs – energy in finding a mate; having to copulate/mate; internal benefits – fewer eggs required; higher success/survival rate for each offspring; external costs – low fertilisation chances; many gametes wasted; high mortality rate in offspring; external benefits – large number of offspring; no mating/copulation costs. **Any 6**

Check-up 38

1 Monogamy – each individual mates only with one other; polygamy/polyandry/polygyny – an individual mates with a number of others/females/males. **2**

2 Polygyny is one male and several females; polyandry is one female and several males. **2**

3 There are sign stimuli, which trigger responses; there are fixed action patterns, which dictate courtship and mating. **2**

Check-up 39

1 Each sex looks different. **1**

2 Males display to attract females to select them for mating. **1**

3 Increased chance of a mate able to raise/feed young; suggests a male with a low parasite count. **2**

4 Male are armed/have armaments/weaponry/examples; fight other males for mating rights. **2**

Answers to exam-style questions

Structured questions

1 a) Sexual dimorphism. **1**

b) Territorial male has best chance of securing breeding rights; satellites gain chances to mate even though they do not hold territory. **2**

c) Satellites attract additional females to the lek. **1**

d) (i) An individual can mate with several partners. **1**

(ii) Cryptic/camouflaged on/at nest. **1**

2 a) (i) Species in which males display to compete for females. **1**

(ii) Female choice is based on criteria not directly related to survival; but which give better mating chances. **2**

(iii) If male fit/vigorous/has low parasite count; better able to raise/feed young so better survival chances for offspring. **2**

b) (i) Rut is male–male rivalry through aggression/fighting; females don't choose, winner secures mating rights. **2**

(ii) Large size; large antlers; much stamina; experienced fighter. **Any 2**

Extended response

3 Compared with r-selected, K-selected have larger body size; longer lifespan; reproduce later in life; reproduce more often; have a smaller number of offspring; have larger offspring; have higher levels of parental care; offspring have higher survival rates; occupy more stable habitats [each statement must contain a comparison]. **Any 8**

4 Courtship is a behaviour adapted to successful reproduction/mating; can be linked with sign stimuli; these are specific aspects of the appearance/behaviour of the different sexes; example given (red underparts of male sticklebacks/lekking behaviour by male black grouse; lekking behaviour described); sexual dimorphism is difference in appearance in females compared with males; females assess male fitness; using honest signals (shown in the display); fitness is a measure of advantageous genes/chances of survival/raising offspring/low parasite count; can be linked to fixed action patterns; which are sequences of behaviour leading to mating often involving ritualised behaviour. **Any 8**

2.5 Parasitism

Answers to check-up questions

Check-up 40

1 Fundamental is occupied in absence of interspecific competition; realised is occupied when interspecific competition is present. **2**

2 Competitive exclusion occurs when competition makes a species extinct locally; resource partitioning allows co-existence by division of resources. **2**

3 Co-evolved/intimate relationship between species; parasite (+)/host (−); reproductive potential of parasite greater than host; narrow/specialised niche; host-specific; many parasites are degenerate. **Any 4**

Check-up 41

1 Sexual reproduction occurs within the definitive host; secondary/intermediate host allows the parasite to complete its life cycle. **2**

2 Actively transmits the parasite to a new host. **1**

3 *Plasmodium* enters human in mosquito bite; moves to liver and red blood cells; gametocytes enter blood; ingested by mosquito during bite; sexual reproduction occurs; stages move to salivary glands. **Any 5**

4 RNA/DNA/nucleic acid; in a protein coat. **2**

5 Nucleic acid enters host; nucleic acid replicated; genes translated to viral protein; assembly of new viruses; viruses leave/burst out of cell. **5**

6 RNA converted to DNA by reverse transcriptase; DNA enters host genome; viral genes expressed to form new viral particles. **3**

Check-up 42

1 Ectoparasites by direct contact/ingestion of infected material; endoparasites (often) by vectors. **2**

2 Overcrowding of hosts; abundance of vectors. **2**

3 Alteration of host foraging/movement/sexual behaviour/habitat choice/anti-predator behaviour [Any 2]; to maximise transmission to new host. **3**

Check-up 43

1 Hydrolytic enzymes in mucus/saliva/tears; destroy bacterial cell walls; low pH secretions of stomach/vagina/sweat glands; denatures cellular proteins of pathogens. **4**

2 Injured cells release signalling molecules; enhanced blood flow to the site; antimicrobial proteins and phagocytes brought to site. **3**

3 Engulf pathogen into vacuole; add digestive enzymes from lysosomes; destroy; digest pathogen. **3**

4 Identify and attach to infected cells; releasing chemicals that lead to cell death; by inducing apoptosis. **3**

Check-up 44

1 Specific antigen–antibody reaction; lymphocyte then produces a clone of identical cells (same antibody). **2**

2 Antibody has a variable region; binds to any one type of antigen. **2**

3 Release of antibodies; induce apoptosis. **2**

4 Higher concentration of antibodies; faster production of antibodies; longer-lasting effects of antibodies. **Any 2**

Check-up 45

1 Mimic host antigens to evade detection; modify host immune response to reduce chances of destruction; variation of antigens during an infection; integrating their genome into host genomes and entering latency. **Any 3**

2 Level of immunity in a population which can prevent an epidemic. **1**

3 Similarities between the host and parasite metabolism. **1**

4 Vaccines contain antigens that trigger an immune response; but are harmless. **2**

5 Difficult to design vaccines because of antigenic variation; some parasites are difficult to culture for vaccine production; overcrowding in LEDCs/refugee camps/war zones/areas after natural disasters; areas with tropical climates, which favour parasites. **4**

Answers to exam-style questions

Structured questions

1 a) A multi-dimensional summary of tolerances and requirements of a species. **1**
  b) W – fundamental niche of species 1; X – fundamental niche of species 2; Y – realised niche of species 1. **3**
  c) Resources are divided between the two species by competitive exclusion. **1**

2 a) (i) Sexual reproduction occurs in the human. **1**
    (ii) It does not actively transmit the parasite. **1**
  b) (i) (Larval) stages are waterborne. **1**
    (ii) Avoid wading in water; use molluscicide/kill snail. **2**
  c) Mimic host antigens; modify host immune response; antigenic variation; latency. **Any 2**

Extended response

3 a) Epithelium forms a barrier; chemical secretions discourage/kill pathogens; phagocytosis engulfs pathogen; natural killer cells kill infected cells; inflammatory response causes vasodilation, which brings defence cells to site. **Any 4**

  b) Cytokines increase blood flow to infection sites; lymphocytes are specific to pathogens; lymphocytes produce clones; lymphocytes recognise foreign antigens; produce antibodies against specific antigens/pathogens; antigen–antibody complex forms; renders pathogen harmless. **Any 6**

4 a) Difficult to design vaccines because of antigenic variation; some parasite difficult to culture for vaccine production; overcrowding in LEDCs/refugee camps/war zones/areas after natural disaster; areas with tropical climates which favour parasites. **4**

  b) Civil engineering projects improve sanitation; co-ordinated vector control; reduction of child mortality; population-wide improvements in child development/intelligence; because individuals have more resources for growth and development. **Any 3**

Answers to practice course assessment

Section 1

1 D	3 D	5 B	7 B	9 A
2 A	4 B	6 C	8 D	10 B

Section 2

1 a) $N = \dfrac{MC}{R}$ **1**
  b) Study population is closed; no individuals die/are born/immigrate/emigrate between visits; marks do not disappear between visits; all individuals have an equal chance of capture; released individuals can mix fully and randomly. **Any 2**
  c) Painting; banding; ringing; surgical implantation; GPS tracker; hair clipping; ear clipping. **Any 1**
  d) Indicator species/to investigate environmental conditions *or* qualities/presence of a pollutant. **1**
  e) Minimise impact on wild species *or* habitats/disturbance to wild populations. **1**

2 a) The change over time in the proportion of individuals in a population differing in one or more inherited traits. **1**

b) (i) Housefly population rises in
       container B. **1**

(ii) Parasite applies selection pressure
     to houseflies; mutations give
     resistance; natural selection favours
     resistant flies. **Any 2**

c) (i) The process by which two or more
       species evolve in response to selection
       pressures imposed by each other. **1**

(ii) Wasp numbers rise and houseflies
     population falls. **1**

3 a) (i) More resources into the egg structure. **1**

(ii) Increased probability of production *or*
     survival of offspring. **1**

b) Smaller; shorter generation time; mature
   more rapidly; reproduce earlier in their
   lifetime; many offspring; limited parental
   care; offspring receive smaller energy input;
   very low offspring survival. **Any 2**

c) (i) Polygyny. **1**

(ii) Increased chances of offspring survival/
     increased reproductive fitness. **1**

4 a) Asexual reproduction in which offspring
      develop from unfertilised eggs. **1**

b) There are more rusts/fungi at lower
   altitudes, which are warmer and
   more humid; at these latitudes
   sexual reproduction is favoured over
   parthenogenesis because it gives the
   variation needed to evolve with parasites. **2**

c) (i) Advantage – preserves successful
       genotypes *or* whole genomes are
       passed on to offspring/rapid/allows
       quick colonisation of new/disturbed
       habitats; disadvantage – offspring not
       varied/identical. **2**

(ii) Mutation. **1**

5 a) Parasite gains and host loses. **1**

b) Humans–where sexual reproduction
   occurs/parasite reaches sexual maturity. **1**

c) Helps the parasite find new hosts/
   achieve transmission/developmental
   stages of parasite/site of asexual
   reproduction. **1**

d) People need to enter water to tend rice
   plants; sewage systems in these areas not
   well developed. **1**

e) Reduce effect of immune system/
   mimic host antigens/antigenic variation/
   integrate genome into host genome *or*
   latency. **1**

f) Degenerate. **1**

6 a) Occurs in gamete/diploid mother cells;
     chromosomes condense; in meiosis
     I homologous chromosomes pair; at
     the equator of the metaphase spindle;
     homologous chromosomes separate/
     segregate; in meiosis II chromosomes align
     at equators; sister chromatids split/move
     apart; four haploid gametes produced;
     independent/random assortment/
     segregation at meiosis I increases variation;
     chiasma form at meiosis I; alleles undergo
     crossing over; recombination increases
     variation. **Any 10**

b) Can be environmental; based on
   temperature/competition/size/infection
   [any 2]; and can change the sex of
   individuals in some species; in reptiles
   incubation temperature can determine
   sex of offspring; can be genetic; sex
   chromosomes in mammals/birds/some
   insects; in mammals XX produces female;
   and XY produces male; SRY allele on
   the Y chromosome; encodes maleness/
   TDF gene; females homogametic/males
   heterogametic. **Any 10**

Area 3 Investigative biology answers

3.1 Scientific principles and process

Answers to check-up questions

Check-up 46

1 Observation; hypothesis formation; experimentation; concluding and evaluating. **4**
2 That there is no statistically significant effect of the independent variable on the dependent variable; easy to test then modify. **2**

Check-up 47

1 Scientific papers; seminars/conferences/symposia. **2**
2 Introduction and abstract; method/procedures; results/analysis; discussion/conclusion; evaluation. **5**
3 Summarises current knowledge/recent findings in a field; written by leading scientists in that field. **2**
4 Replacement; reduction; refinement. **3**

Answers to exam-style questions

Structured question

1 Observation of nature; hypothesis formation; experimental design; experimentation; gathering of data; analysis of data; evaluating and concluding; publication. **5**
[lose 1 for each stage in an incorrect position]

Extended response

2 Replacement; is the use of cells/tissue samples instead of whole organisms; reduction is the minimising of the number of animals needed for a study/altering the statistical methods to allow for lower number of animals; refinement; involves alteration to method to minimise harm; including consideration of welfare issues such as food/accommodation. **6**

3.2a Experimentation – experimental skills

Answers to check-up questions

Check-up 48

1 Validity concerns the control of variables; so that conclusions regarding variables are fair; reliability concerns consistency of results; after independent replication of experiments. **4**
2 Accuracy is concerned with how close to a true value a measurement is; precision is concerned with how close replicated measurements are to each other. **2**
3 Continuous measured in infinite numerical values; discrete limited to whole number values. **2**
4 *In vitro* procedures are controlled in a laboratory outside of a living organism; they are controllable/reliable/cheap; *in vivo* procedures are carried out with whole organisms or in nature; they simulate real life and can test chronic effects. **4**
5 Negative control highlights the role of the independent variable in affecting the dependent variable; positive control proves that the system/procedure can detect positive results when they occur. **2**
6 Placebo is a treatment that does not contain the independent variable/drug/vaccine; an effect on the dependent variable because of patient expectations. **2**

Check-up 49

1 Gender; colour morph; others. **Any 2**
2 Simple experiments have one independent variable; multifactorial have more than one. **2**
3 Removes/reduces effects of confounding variables; that can't be controlled and may affect the dependent variable. **2**

Check-up 50

1 Allow procedures/techniques to be refined; highlight the most appropriate values for independent variables/numbers of replicates. **2**

2 Same mean; degree of variation/standard
   deviation as the whole population.   **2**
3 Systematic is at regular intervals; stratified
   divides the population into groups then
   samples within the groups.   **2**

Check-up 51

1 Natural variation in the biological material;
   sample size.   **2**
2 Repeats are within the same experiment/
   procedure; independent replicates are done
   separately by a different investigator/in a
   different laboratory/at a different time.   **2**

Answers to exam-style questions

Structured questions

1 a) (i) Potato variety.   **1**
      (ii) Temperature/pH **or** age/size of potato. **1**
   b) (i) Designed to show that the procedure
          can detect a positive result.   **1**
      (ii) Set of replicated without G1P **or** with
          no enzyme source.   **1**
   c) So that any starch detected had been
      produced in the experiment/*de novo*.   **1**
   d) (i) Divide the potatoes into age/size groups;
          help to remove potential for
          confounding variables.   **2**
      (ii) Method does not detail number of
          tubers used **or** reliability low because
          only one extract per tuber was used **or**
          reliability could be improved by using
          more extracts.   **1**
2 a) Concentration of ATP solution.   **1**
   b) (i) Age of meat/thickness of strip/other
      relevant answer.   **1**
      (ii) Muscle protein may deteriorate with age/
          ATP diffusion slower in thicker strips. **1**
   c) Not reliable; no independent replication/
      experiment carried out only once.   **2**
   d) Negative control.   **1**
   e) Might have prevented a representative
      sample being taken/sample may not have
      been truly random.   **1**

Extended response

3 Sample should have same mean as
   population; same degree of variation

about the mean as population; sample size
bigger in more variable populations/greater
reliability with larger/more samples; random/
all individuals have equal chance of being
selected/avoid selection bias; systematic –
individuals selected at regular intervals;
stratified – population divided into categories
and sampled proportionately.   **Any 4**
4 a) More than one independent variable; for
      example, fertilisers with several elements
      present; more difficult to analyse;
      confounding variable present.   **Any 3**
   b) Used in field trials; where confounding
      variable cannot be controlled; treatment
      and control groups randomised within the
      blocks; minimises the effect of confounding
      variables; replication built in through
      repeated blocks.   **Any 4**

3.2b Experimentation – data handling skills

Answers to check-up questions

Check-up 52

1 Quantitative has numerical values; qualitative
   is descriptive; ranked is transformed to
   categories.   **3**
2 Quantitative – line graph; qualitative – bar/pie
   chart; ranked – bar chart.   **3**

Check-up 53

1 Mean – arithmetic average; median – middle
   value; mode – most common value.   **3**
2 Data set with some extreme outlying values;
   data that is descriptive.   **2**
3 Its variation; in either direction from
   the mean.   **2**

Check-up 54

1 No significant difference at day 0; error bars
   don't overlap after day 0; so differences could
   be significant.   **3**
2 a) Growth of males and females not significant
      when dominant individual present.   **1**
   b) Growth of males significantly greater; when
      no dominant individual present.   **2**

Check-up 55

1 Comparison between data sets; data in quartiles/upper and lower quartile/interquartile range; median. **3**

2 Range; standard deviation; standard error of the mean; confidence limits. **Any 3**

Check-up 56

1 Clustered along a line of best fit that rises across the plot. **2**

2 The more clustered/fewer outliers, the stronger the correlation. **1**

3 Presence of potentially confounding variables. **2**

Answers to exam-style questions

Structured questions

1 a) (i) Parasite population increases following a grouse population increase, causing the grouse population to decrease again. **1**

   (ii) Weather/predators/other parasites/competition. **1**

   b) Treated birds increase more in mass; because they do not have to support the parasite. **2**

   c) (i) Treated birds raise more chicks over the following three years; adult birds are in better condition/can obtain more food/resist disease better/escape predators better. **2**

   (ii) Number of chicks produced by treated birds also varies from year to year. **1**

   d) (i) As the percentage of treated birds increases, number of grouse increases; treated birds don't pass parasite out in faeces, so less larvae on heather **or** treated birds lose less energy to parasites so breed better. **2**

   (ii) Immigration from other areas; stages of parasite remain on heather stems. **2**

2 a) Rainfall; temperature; invertebrate food availability; other relevant answer. **Any 2**

   b) (i) Correlation – pH related to date of first egg may be indirect and be caused by another variable such as rainfall/pollution. **1**

      OR

Causation – pH is directly related to date of first egg as shown by a line of best fit through the data points. **1**

   (ii) Between pH 5.0 and 6.0 there is a clear relationship between the pH and date of first egg; between pH 7.0 and 8.0 no clear trend/relationship can be seen between pH and date of first egg. **2**

   c) Validity – effect of monitoring/disturbance on dipper behaviour; bias in the selection of nest to study, such as the remoteness of the nest sites. **1**
   Reliability – number of birds studied; seasons involved; areas of the country covered; other relevant answer. **1**

   d) At pH values below 6.0, stoneflies make up the highest percentage of the diet; at pH values above 6.0, mayflies make up the highest percentage of the diet. **2**

   e) Higher percentage/approximately 50% of their diet is made up of mayflies; and a low percentage/approximately 5–10% of stoneflies. **2**

3 a) (i) More likely to reveal nests than a random search. **1**

   (ii) Minimise selection bias that may affect a non-random sample. **1**

   b) That feather mite infestation does not affect the reproductive success of crested tits. **1**

   c) As the number of feather mites increases, the breeding success of the birds decreases. **1**

   d) (i) As the size of the oil gland increases, the number of feather mites decreases. **1**

   (ii) Ensure that the oil gland measurements were made at the same time of day. **1**

   e) Obtain correct licences; ensure as few individuals as possible are disturbed; keep study sites confidential; avoid making access to sites easier for predators; do pilot studies to quantify effects of disturbance. **Any 1**

Extended response

4 Mean is arithmetic average; mode is commonest value; median is middle value; median better when a small number of extreme values present in data; mode better when data is descriptive/not numerical. **5**

5   Scatter plots give lines of best fit; rising line indicates positive correlation; falling line indicates negative; slope of line indicates strength; cluster of plots indicates strength.   **5**

3.3 Reporting and critical evaluation of biological research

Answers to check-up questions

Check-up 57

1   Underlying biological context; hypothesis; justification of selections of procedures.   **3**

2   Enough detail for an independent repeat to be carried out.   **1**

3   Appropriate graphs/table of data; appropriate statistics; use of error bars on graphs; evidence of the consideration of outliers and anomalous results.   **Any 3**

4   Conclusions related to aim (and hypothesis); evaluations of method/conclusions; awareness of the contribution of the research to knowledge in science.   **3**

5   Lab book should be like a diary; contains notes and results on pilot studies; decisions about method and procedures; draft results table; processing such as statistical calculations; notes on outliers/anomalous results; draft conclusion; references; notes on evaluations.   **Any 5**

Answers to exam-style questions

Structured question

1   a)   A brief summary of the aims; and findings of the work.   **2**

b)   References embedded in text; with full reference in reference list.   **2**

c)   Results data that fall well outside the range of the data in general; may be discarded.   **2**

d)   Result of failure to take random samples.   **1**

Extended response

2   Observation; construction/revision of hypothesis; experimentation; gathering/ recording of data; analysis; concluding; evaluating; reporting of findings; new/modified hypothesis.   **Any 8**

Answers to practice course assessment

Section 1

1   D   3   C   5   A
2   D   4   D

Section 2

1   a)   Compared with the control, drug treatment increased the adiponectin concentration in the blood more than lifestyle changes.   **1**

b)   (i)   Their health and well-being; that they have been given all the appropriate information; that they have given signed consent; that the terms of the study are legal and risk assessed.   **Any 1**

(ii)   Because of the natural variability of the population.   **1**

2   a)   Independent – frequency of fanning bouts; dependent – hatching success.   **2**

b)   Increased fanning gives eggs more oxygen; this increases their chances of hatching.   **2**

c)   More frequent fanning; lower bout duration.   **2**

3   a)   Faecal sample collected remotely from the habitat.   **1**

b)   (i)   As vole numbers/abundance increase, the percentage of grouse remains in marten scat samples decreases.   **1**

(ii)   The abundance of other foods; climatic conditions affecting hunting; number of pine martens; number of scat samples; time of sampling.   **Any 1**

(iii)   A data point that is significantly far away from the trend in the data of a random sample of the population.   **1**

(iv)   Increase the sample size/collect more data.   **1**

4   a)   Increasing the number of boxes increases the number of pairs; numbers of blue tits increase more than number of great tits.   **2**

b) (i) As the number of nest boxes increases, the percentage of blue tits using them increases rapidly then more slowly. **1**

(ii) The error bars overlap; however, the results are significant because overlap between values for widely spaced values of the independent variable do not overlap significantly. **1**

c) 125 / 25 = 5 boxes per hectare; c. 62% blue tits so c. 38% great tits. **2**

5 a) Help plan/modify/evaluate/develop/ practice procedures; assess validity of experimental design; check effectiveness of techniques; establish appropriate range of values for the independent variable; identify suitable number of repeat measurements/ replicates required to give representative values; assess timescale to complete investigation. **Any 5**

b) Refine methods to minimise suffering; replace animals with alternative techniques; reduce the number of animals used to the minimum needed to provide meaningful results; to avoid harm; humans – obtain informed consent; evaluate harm/risk assess; the right to withdraw data; confidentiality. **Any 5**